THE GENOME FACTOR

THE GENOME FACTOR

What the Social Genomics Revolution
Reveals about Ourselves, Our History,
and the Future

**DALTON CONLEY
AND JASON FLETCHER**

PRINCETON UNIVERSITY PRESS
Princeton and Oxford

Library of Congress Cataloging-in-Publication Data

Names: Conley, Dalton, 1969– author. | Fletcher, Jason, author.
Title: The genome factor : what the social genomics revolution
    reveals about ourselves, our history, and the future / Dalton
    Conley and Jason Fletcher.
Description: Princeton : Princeton University Press, [2017] |
    Includes index.
Identifiers: LCCN 2016036571 | ISBN 9780691164748
    (hardback)
Subjects: LCSH: Genomics--Social aspects. | Heredity,
    Human—Social aspects. | Equality. | BISAC: SOCIAL
    SCIENCE / Sociology / General. | SCIENCE / Life Sciences /
    Genetics & Genomics. | SOCIAL SCIENCE / Disease &
    Health Issues.
Classification: LCC QH438.7 .C656 2017 | DDC 576.5—dc23
    LC record available at https://lccn.loc.gov/2016036571

British Library Cataloging-in-Publication Data is available

This book has been composed in
Sabon Next LT Pro & Montserrat.

Printed on acid-free paper. ∞

Printed in the United States of America

1 3 5 7 9 10 8 6 4 2

CONTENTS

ACKNOWLEDGMENTS

At least three key programs helped set up the foundation for our collaboration for this book. The first is the Integrating Genetics and Social Science Conference (IGSS), directed by our friend and collaborator Jason Boardman. One of our early interactions was Conley dismantling Fletcher's paper at the first annual conference, which is now in its seventh year at the University of Colorado's Institute for Behavioral Science. This conference has formed the epicenter of much of the science we highlight in the book, and we thank Jason Boardman, Jane Menken, Richard Jessor as well as the staff and funders of the conference, Population Association of America, the NICHD, IBS, and Colorado Population Center, among others, for such an invigorating and engaging (and continuing) set of events. The second is the Robert Wood Johnson Foundation Health & Society Scholars Program at Columbia University. Fletcher spent 2010–2012 in the program, which allowed us to discuss joint work that set some of the groundwork for the book. Fletcher also deeply thanks Peter Bearman, Bruce Link, and Zoe Donaldson for advice, support, and engagement during this time as well as former colleagues at Yale who cultivated this research direction, especially Paul Cleary, Joel Gelernter, and Mark Schlesinger. Conley is grateful to the NYU biology department for allowing him to go back to school to learn the difference between siRNA and miRNA and piRNA. And thanks also to the NYU administration that countenanced such an unusual arrangement—especially to Provost David McLaughlin.

Along the way, we have benefited from advice and support from a large number of colleagues and collaborators. Our friends and collaborators in this domain of research include but surely are not limited to (alphabetically): Dan Belsky, Daniel Benjamin, Jason Boardman, Richard Bonneau, David Cesarini, Justin Cook, Christopher Dawes, Ben Domingue, Kathleen Mullan Harris, Phillip Koellinger, Thomas Laidley, Steve Lehrer, Patrick Magnusson, Matthew McQueen, Michael Purugganan, Emily Rauscher, Niels Rietveld, Lauren Schmitz, and Mark Siegal. They have all substantially enhanced our research efforts as well as the wider field. We also thank Jason Boardman, Justin Cook, Mitchell Duneier, Angela Forgues, Joel Gelernter, Joel Han, Ryne Marksteiner, Ann Morning, Jessica Polos, Matthew Salganik, and Maria Serakos for their extensive comments on various chapters of the book. In the Princeton Sociology Department, Amanda Rowe copyedited several versions of the manuscript, improving it with each pass.

We have been generously funded over the last years by a variety of organizations.

Conley gratefully acknowledges the Russell Sage Foundation for both his time as a visiting scholar in residence there as well as the subsequent awarding of a research grant that supported some of this work (Grant # 83-15-29: "GxE and Health Inequality across the Life Course"). Conley also thanks the John Simon Guggenheim Foundation for its Individual Fellowship ("In Search of Missing Heritability"). Finally, he would like to acknowledge the National Science Foundation, which supported his second Ph.D. studies in the form of the Alan T. Waterman Award (SES-0540543). Internal research support from both New York University and Princeton University also made this work possible. Visiting stints at the University of Auckland, Bielefeld University, Yale University's Center on Inequality and the Life Course and at the Institute for Behavioral Science at the University of Colorado at Boulder were all helpful opportunities for Conley to develop the ideas herein. In particular, he would like to thank hosts Richard Breen (Yale, now Oxford); Peter Davis (Auckland); Martin Diewald (Bielefeld) and Jason Boardman (Colorado).

Fletcher gratefully acknowledges the Robert Wood Johnson Foundation Health & Society Scholars Program and the William T. Grant Foundation Scholars Program for career development support. Fletcher is particularly grateful for the advice from Adam Gamoran, Richard Murname, David Deming, Joshua Brown, Patrick Sharkey, and Jelena Obradovic at the Scholars Retreat workshop. Fletcher also is grateful for research support and generous colleagues from the La Follette School of Public Affairs, Department of Sociology, Center for Demography and Ecology, Center for Demography of Health and Aging, and Institute for Research on Poverty at the University of Wisconsin-Madison.

We also have benefited from numerous kind souls at Princeton University Press who have expertly helped us at every step. Thanks to Eric Schwartz who acquired the book before decamping for Columbia University Press. Meagan Levinson picked up where Schwartz left off by providing exceptional guidance and a thoughtful and patient ear to keep us on track and our audience in mind. Gail Schmitt provided excellent copyediting. Leslie Grundfest and Karen Carter were very efficient production editors. If you are reading this, it is probably due to the efforts of Julia Haav and Caroline Priday in the PUP publicity department. Conley, in particular, wants to thank Peter Dougherty, the director of the press, who first wanted to publish him almost 20 years earlier, when Conley was just out of graduate school (the first time) and Dougherty was the sociology editor for the press. Lastly, the entire staff helped smooth out all the wrinkles.

Last, but not least, we both have wonderful, supportive, loving families to thank. Conley would like to express his gratitude to his kids E and Yo for teaching him the power of genetics through their differences and for providing the perfect audience for him to practice explaining his ideas over take-out dinners; his parents, Steve and Ellen, for bequeathing much more than their DNA; and to his partner, Tea Temim, who challenges his assumptions (and math), always. Fletcher owes an incalculable (but never tallied) debt to Erika, Anna, and Isaac for the gift of time to concentrate on his first book project and their encouragement; to Phil, Paula, Jim, Cindy, Ann, and Justin for active engagement and interest in what he has been up to and their unconditional support.

And, finally, thanks to the untold generations before us who have bequeathed both their genes and their culture, even if that dual-patrimony makes our statistical models all the more complicated.

THE GENOME FACTOR

CHAPTER 1

MOLECULAR ME

WELCOME TO THE COMING
SOCIAL GENOMICS REVOLUTION

The social genomics revolution is now upon us. In a little more than one hundred years, a gene—a discrete unit of heredity—has been transformed from a vague notion among a small group of scientists to almost a ubiquitous news item and a consumer commodity. The black box of the genome has been cracked open by inexpensive DNA genotyping platforms that for a mere hundred dollars allow one to measure around a million of the base-pair differences between individuals. We can now all drink from a firehose of new data and research findings aimed at describing the underlying genetic architecture of human health and well-being. In addition to the large number of biologists and medical scientists working in this vein, a small but growing group of sociologists, political scientists, and economists have joined forces with statistical geneticists to make serious arguments about the role of genes in the broader domain of human social dynamics and inequalities.

A note on how to read this book: We wrote the main text with the general reader in mind. Endnotes and appendices to each chapter are meant for those who want to dig deeper into the technical details and scientific minutia.

This component of the genomics revolution is somewhat unusual in that it follows a long history during which geneticists and social scientists have been unwilling to work with each other on important scientific issues. Indeed, for most of history since Darwin penned *The Descent of Man*—his 1871 follow-up to *The Origin of Species* that focused on the evolution of human differences—there has been a contentious relationship between biologists and social scientists, with many examples of interdisciplinary exchanges having devastating impacts on society. There was Herbert Spencer, who applied natural selection as a metaphor to human society leading to the ideology of "social Darwinism," which justified inaction in addressing all manner of social ills and inequities. And there was Darwin's cousin, Francis Galton, who pioneered eugenics. Even Darwin himself became embroiled in a debate about whether blacks and whites constituted separate species.[1]

Part of the reason for trepidation by social scientists when it comes to the examination of genetics as it relates to human behavior is that it is commonly assumed that the answers obtained by looking at genetics are deterministic ones, which is at odds with much of the social science enterprise. Furthermore, to the extent that genes explain any social phenomenon, this fact "naturalizes" inequalities in that outcome. In other words, the degree to which IQ (or height) differences between people are genetically influenced might anchor beliefs that these outcomes are innate and thus immutable. If these "natural" differences between people extend to outcomes such as education levels or earnings (i.e., income inequality), such inequalities may be argued to be natural, unchangeable, and thus outside the realm of policy intervention—"naturalizing" or rationalizing inequity.[2]

One important example of the naturalization of social or economic inequality can be found in the best-selling book, *The Bell Curve: Intelligence and Class Structure in American Life*, in which Richard Herrnstein and Charles Murray argued that thanks to meritocracy, class stratification today is based on innate (i.e., genetic) endowment.[3] By selectively breeding with others of similar genetic stock, parents reinforce their offsprings' advantages or disadvantages. According to these authors, social policy to promote equal opportunity is counterproductive since individuals have reached the level of social status best suited to their native abilities. This conclusion is the nightmare of progressive social scientists and the primary reason most of them avoid genetic data.[4]

We, instead, head straight into this domain of inquiry with eyes wide open. The genetics of inequality is, in fact, a major theme of this book. Specifically, we ask how integrating molecular genetic information into social scientific inquiry informs debates about inequality and socioeconomic attainment (of individuals as well as of entire nations). We argue that there are three main ways.

First, by dealing directly with the contention that innate, inherited differences are the primary engine of social inequality, the integration of genetic markers shows the residual social inequalities in stark relief. By actively accounting for the portion of IQ, education, or income that is the result of genes, we can see more clearly the inequities in environmental inputs and their effects on individuals' chances in the game of life.

Second, we show how genotype acts as a prism, refracting the white light of average effects into a rainbow of clearly observable differential effects and outcomes. Our intuition is that genotype is a tool that will help us understand why, for example, childhood poverty wreaks havoc on some individuals whereas others are resilient to such traumas. Or, by explicitly integrating genetic information into social science, we can go from the adage that a gene for aggression lands you in jail if you are from the ghetto but in the boardroom if you are to the manor born, to a scientific research agenda showing how environmental and genetic effects mutually depend upon each other.

Third, we think that public policy will have to deal with this new information as lay citizens get a hold of their own (and others') genetic data. There is much written about privacy, "genoism" (genetic discrimination by the likes of insurance companies), and personalized medicine. We shift attention from these topics to tackle the more traditional social policy domains of education, income support, economic development, and labor markets and interrogate the implications of genotype for those realms. We argue that, contrary to claims in *The Bell Curve*, we have not yet become a genotocracy (a society ruled by the genetically advantaged), but that would not be such a far-fetched possibility once those with power and resources start to control their own genetic information and use it to selectively breed themselves. The social genomics revolution means that new forms of inequality may emerge based not only on genotype but also on whether individuals know their genotypes (and the genotypes of those around them) and can act on that information.

This revolution, however, ran into some stumbling blocks early. A few initial successes in finding "the gene for X" (such as age-related macular degeneration) produced false hope and optimism that uncovering the genetic basis for much common disease, and even socioeconomic outcomes, was within reach.[5] These successes were followed by failures—both null findings for some outcomes and (in hindsight) false insights into others. The lessons of statistical research—the need for adequate sample sizes and clear hypotheses—were learned anew the hard way by this emerging field. The result of these and other challenges was that measured genetic markers did not seem to explain the amount of variation in outcomes—ranging from schizophrenia to height to IQ—that was predicted to be due to genetics according to the prior generation of studies. For example, some earlier heritability estimates for schizophrenia reached higher than 80 percent,[6] yet some of the best studies that used DNA data produced heritability estimates closer to 3 percent—leading some scientists to dub this problem "missing heritability."[7] (We discuss heritability in chapter 2 and the mystery of missing heritability in chapter 3.)

Recently, geneticists began to reevaluate the reasons that this heritability was missing. Perhaps because the earlier studies used twins and family relationship data instead of actual DNA data, the role of genes was overestimated to begin with. Or maybe researchers had been looking in the wrong places in the genome. Genotyping companies have traditionally focused on measuring common genetic variants rather than rare ones. Perhaps there are still genes for "X" (e.g., schizophrenia) that remain undiscovered because they are rare in the population and not typically measured in the DNA data that are available. Slowly but surely, statistical geneticists have made significant progress in solving this "missing heritability" mystery by using a range of newly developed tools.

The current and more widely accepted answer to this missingness relies on a paradigm shift—moving from a mindset of a "gene for X" to one of "many variants with small effects." Instead of a single important genetic variant (or allele), there are often hundreds or thousands that contribute to variation in a given outcome. But the "small effects" aspect of this paradigm has called for ever-larger data sets to find the needles in the genomic haystack because the needles are now thought to be much smaller than originally suspected.

Along with this shift, more national surveys have been asking respondents to spit into a cup, adding genotype data to the rich tapestry of social variables that economists, sociologists, and political scientists had previously used. It seems that genetics has finally gained a foothold in social science. And why not? Why should we be afraid of additional data that may help scientists better understand patterns of human behavior, enhance individuals' self-understanding, and design optimal public policy? Why be apprehensive, especially when the answers we get from carefully peering into the black box are not always—or even often—the kind that crudely reify existing inequalities, assumptions, and policies? As it turns out, new discoveries made by adding genetic data to social science are overturning many of our assumptions. For example, were Herrnstein and Murray correct in *The Bell Curve* when they argued that meritocracy has perversely resulted in more intransigent inequalities today because we are now sorted by genetic ability? Probably not. The data show that the magic of sexual reproduction and other genetic processes do as much (if not more) to upset existing inequalities (i.e., create social mobility) than they do to reinforce social reproduction. This molecular shake-up results from two main forces. First, while spouses are somewhat correlated in their genetic signatures, that correlation is weak enough to suggest that there is much dilution and resetting that occurs when someone at one end of the genetic distribution seeks a mate with which to reproduce: her offspring are likely going to revert to the mean. Second, to the extent that outcomes depend on particular combinations of alleles (known as dominance or epistasis), mating disrupts extreme genotypes. We can describe this magic in terms that card players understand. In a common version of stud poker called Pass the Trash, players are dealt seven cards but then must pass three of them to their neighbor (and receive three from their other neighbor). You can be dealt a royal flush at first (i.e., be born with great genes), but once you have to join your cards (DNA) to someone else's, all your advantages can be reshuffled (i.e., your kids have no particular genetic advantage). This reshuffling can have big effects even if you have an advantaged neighbor (i.e., spouse) with whom to trade cards (DNA). You might lose your diamond royal flush when your queen of diamonds (which you were forced to pass to your neighbor) is replaced by his queen of hearts. In these two ways, then, the magic of sexual reproduction acts to reshuffle the deck of cards in each generation, preventing stable genocratic castes from emerging despite the well-observed affinity of like to marry like (we discuss trends in this kind of sorting along genetic and nongenetic lines in chapter 4).

Or take the case of the most sensitive issue of all when it comes to human genetics: race. What if we said that both of the authors, European mutts of mixed white ancestry, are genetically more similar to, say, a Mongolian, than the Luo tribe of Kenya is to the Kikuyu ethnic group of Kenya? You may not believe me (or your eyes), but it is true. Looks can deceive when it comes to racial classification. In fact, the entire community of non-African (and non–African American) human beings collectively can display the same level of genetic similarity as the population of a single region of sub-Saharan Africa (namely, the Rift Valley, where humans originated and which remains the deepest wellspring of human genetic diversity). That result is because the population that first left Africa to populate the rest of the world was at one point as small as 2,000 individuals, creating a bottleneck that filtered out much genetic diversity. It is not clear whether that 2,000 figure was the result of mass starvation from an original intrepid caravan of 10,000 or whether the 2,000 was the total number of adventurous eccentrics who thought it a good idea to cross the Sahara or the Red Sea in search of their fortune. Either way, the result of such a bottleneck is to reduce diversity. So, while it may be obvious why the new (mostly humorous) iPhone app "Wait! Don't Fuck Her, She's Your Cousin" has cropped up in an island community like Iceland, the real question is why we all have not downloaded it. Indeed, new research by us and others shows that the typical marriage in the United States is between people who are the genetic equivalent of second cousins (see chapter 4).

The upshot is that our very notion of race—often based on "natural" physiognomic differences such as eye shape, hair type, or skin tone—is, for lack of a better word, just plain *wrong* in genetic terms, as we discuss in chapter 5. Indeed, thanks to the window onto human history that genomic analysis provides, we are now able to resolve all sorts of mysteries of human prehistory, ranging from whether we did it with Neanderthals (yes, if you are of European or Asian descent) to how fertile Genghis Khan was (very) to when and how humans populated New Zealand (not all that long ago, it turns out). From a contemporary perspective, our newfound understanding of human migration and genetic segregation can explain some conundrums. This book will tiptoe through the minefield of race and genetics to confront unspoken beliefs head on. What does a reconstituted understanding of race look like in light of surprising genetic information? How should policy deal with this stubborn concept?

It is not just racial difference (and sameness) that is informed by genetic analysis. The genetic prism can help us understand the rise and fall of entire nations. In chapter 6, we step back and take a macro view of how genetic theories and findings are infiltrating a broader class of global questions; for example, a fundamental domain of inquiry in macroeconomics is why some countries have thrived while others have stagnated during the past several hundred years. There is a long-standing set of hypotheses suggesting that long-ago events and circumstances had lasting impacts that have shaped the tremendous differences in economic success and growth that exist around the world today. Everything, from the north-south versus east-west continental orientations to the shapes carved into the Earth during the last ice age, seems to matter for the wealth of nations. A newcomer to this discussion of "deep determinants" of economic success is population genetics. A new breed of macroeconomists has posited that genetic diversity within countries is a key to development. In 2013, Quamrul Ashraf and Oded Galor published a paper claiming that a "Goldilocks" level of genetic diversity within countries might lead to higher incomes and better growth trajectories.[8] The authors discuss the observation that there have been many societies with low diversity (e.g., Native American civilizations) and populations with high diversity (e.g., many sub-Saharan African populations) that have experienced pallid economic growth, whereas many societies with intermediate diversity ("just right"—i.e., European and Asian populations) have been conducive to development in the precolonial as well as the modern eras. The researchers hypothesize that the Goldilocks advantage results from disadvantages at the extremes of genetic diversity—very low levels of diversity lead to a lack of innovation because everyone is the same, but populations with very high levels of diversity cooperate less because no one is the same.

In addition to calculating the Goldilocks level of genetic diversity with respect to economic development, economists have considered the role of population genetics as it interacts with environmental resources to affect growth patterns across countries. Justin Cook has shown that populations in early human history with the (genetic) ability to digest milk after weaning were conferred large advantages in population density around 1500 CE.[9] Because other studies have shown that historical differences in economic development have been remarkably persistent into the present day, the implication of Cook's study is that (relatively) small changes in the genome at the right time and in the right place (i.e., during the Neolithic Revolution in areas able to domesticate cattle) can lead to large, persistent, and accumulating differences in economic development across countries. But these genes confer advantages only when they occur in environments that have the ability to foster agriculture. With no cows, goats, or other domesticable mammals, the gene confers no population advantage.

In chapter 7, we further expand our discussion by more explicitly bringing the environment back into the conversation and discuss the many complications that emerge when one takes an integrative view of genes and environments. Indeed, genetics and environmental factors seem to interact with one another in a complex, dynamic feedback loop that further explains several aspects of the behavior of humans and entire societies. One strand of this research asks whether genetic factors may predispose people to be particularly sensitive to environmental variation. The idea is that some people are orchids and will thrive (or wilt) based on environmental enrichment (or environmental disadvantage), and other people are dandelions and are relatively immune to their surroundings, for better or worse. If we can tell the difference between orchids and dandelions at a young age, should we use this information in assigning them to their teachers, classrooms, after-school programs, and the like?

This question is evoked by the very mixed success rate that social policies have had up to the present time. Some interventions are successful for some people or during some periods of time, but not others. New evidence merging genetics and public policy has started to uncover *why* we see such different impacts of the same policy for different people and how future policies might be adjusted accordingly, thus extending the concept of "personalized medicine" to allow for "personalized policy." If for genetic reasons some people do not respond to health policies, such as taxes on sugary drinks or cigarettes, should we still make them pay the tax?

Evidence has also suggested that some educational interventions have greater or lesser effects depending on the targeted students' genotypes. Should some students be targeted for future interventions while the resources are diverted from other students? What if we find out that the Earned Income Tax Credit spurs low-wage workers to labor more or less depending on their genes? Whether we want to go down the path toward genetically personalized policy is one of the topics of our conclusion.

While the implications of the genetic revolution for social policy are being debated, a second implication of the social genomics revolution has been the democratization of genetic information. Nowadays, scientists can no longer conceal this information from the public or shape the information to fit their research agendas or political predilections. The cost of human genotyping is dropping faster than Moore's law predicts for the price of computer chips. As a result,

Americans are genotyping themselves in record numbers using consumer services like 23andme, Navigenics, and Knome. And those persons are acting on the information they receive, sometimes even when that information only weakly indicates a predisposition: they are asking their doctor if a procedure or test might be right for them. This is a new direction of consumer-driven medicine, in which patients no longer need to get their medical data from their doctors. A longer-term trend starting with home pregnancy tests and diabetic blood sugar tests has shifted abruptly to assessing immutable health traits—your genotype. Some individuals (such as Angelina Jolie) are seeking preventive bilateral mastectomies to reduce their risk of breast cancer. Couples learn about their status as carriers for a hereditary form of hearing loss and make family-planning decisions based on the health of their unconceived children. In our epilogue we outline one dystopian scenario of genotocracy that could emerge from this brave new world of self-directed eugenics—or what the biologist Lee Silver calls reprogenetics.[10]

In short, this book interrogates how new findings from genetics inform our understanding of social inequalities. We survey the literature (including much of our own research), synthesize findings, highlight problems, put forth new hypotheses, and stir in a lot of guesswork. We show how new genetic discoveries may be both disruptive as well as transformational. Indeed, these discoveries often overturn our pre-existing notions of social processes. They also show how limited our current understanding of genetic effects is and that our attempts to cleanly separate genetic and environmental influences face huge challenges. But rather than wallowing in this perceived intractability, we pivot toward a richer integration of social science with genetics to enhance both fields of study. The potential implications are far-reaching. Will we extend ongoing efforts in personalized medicine to create personalized policy? Will we use our new molecular understandings of human difference to fundamentally transform how we view and use race/ethnicity categories around the world? How will we insure that these new discoveries do not further widen social inequalities, as the rich take advantage of genetic information at a faster pace than the poor? We discuss these and many other issues in the pages that follow.

While progress in this burgeoning field is happening at a break-neck pace (at least as compared to typical social science), we feel we must also note that it is still early days. We think the nature-nurture debate that dominated the twentieth century and a portion of the nineteenth is now over, as evidenced by the proliferation of social science surveys that now collect genetic data. That said, the tools for integrating these data are only just being developed and suffer from many limitations that may or may not be overcome in the next de-cade or two. Sample sizes are still too small. Biological mechanisms are hard to pin down. And social systems have a tricky way of learn-ing and adapting, thereby undermining the stability of findings. De-spite these challenges, we think this is an exciting new field worth sharing with you. We hope you agree. The genetics revolution may be well underway, but the social genomics revolution is just getting started.

CHAPTER 2

THE DURABILITY
OF HERITABILITY

GENES AND INEQUALITY

Since the Victorian-era heyday of Francis Galton—the statistician cousin of Charles Darwin and coiner of the word "eugenics"—the heritability of social traits has been the untouchable issue of social science. Some fear that it is a short step from the assertion that complex social behaviors such as criminality or intelligence are highly heritable (as Galton argued in *Hereditary Genius*) to a public policy that includes eugenics. This is not an irrational concern. After all, the very raison d'être for calculating heritability has been to aid breeders. Heritability tells us how quickly a given trait (say, the daily milk or egg yield from livestock) can be changed in the population through selective breeding. It is literally the critical calculation for animal and plant breeders who want to maximize returns on their investments. So why would the rest of us want to calculate heritability if not for eugenics?

Despite the scary, potential applications of heritability calculations to human breeding, some social and behavioral scientists have persisted in trying to estimate the heritability of traits ranging from personality to alcohol consumption to political partisanship. These scholars—mostly psychologists and a smattering of economists—have presented findings that approximately half of the variation in socioeconomic outcomes is due to our genes. That was a radical claim, especially when it was first asserted, in the late 1960s and 1970s, a time when the pendulum had swung to its apogee on the nurture, or environmental, side of the nature-versus-nurture debate. For instance, in the 1970s the notion that sex/gender was entirely socially constructed was fairly accepted (or at least acceptable). It was also in the 1970s that Dr. John Money of Johns Hopkins Medical School touted his sex reassignment surgery for babies with nonnormative genitalia. According to his logic, because sex (biology) was actually subordinate to the environment, any baby—a blank slate of sorts—could be made to fit into any category if treated as belonging to that category from birth. As it turns out, Money's star patient, David Reimer, endured a miserable childhood after being given female genitalia and eventually killed himself.[1]

It was during this heyday of nurture ascendance that—after a quiescent period post–World War II—Arthur Jensen wrote a letter to *Science* claiming that IQ differences were largely explained by genetic variation.[2] A few years later—the economist Paul Taubman "showed" that income was largely determined by genes.[3] These assertions were considered to be heretical ideas during the Age of Aquarius, and other scholars immediately struck back. Most notable was the critique from the economist Arthur Goldberger, who demonstrated the unreliability of Taubman's numbers by showing mathematically that such estimates of heritability could be driven significantly up or down based on assumptions about the extent to which genetic and environmental differences overlapped.[4] (We discuss this critical issue in depth in the next section.)

When the psychologist Hans Eysenck praised Taubman's income study in the press, proposing that the results suggested a limited role for policy in reducing poverty, Goldberger got personal: "A powerful intellect was at work," he wrote with ad hominem sarcasm, which is generally taboo in academic debate. "In the same vein, if it were shown that a large proportion of the variance in eyesight were due to genetic causes, then the Royal Commission on the Distribution of Eyeglasses might as well pack up. And if it were shown that most of the variation in rainfall is due to natural causes, then the Royal

Commission on the Distribution of Umbrellas could pack up, too."[5] The conclusion of mainstream left-leaning academics was that not only was heritability probably fundamentally incalculable, it was also worthless from the point of view of public policy (other than for eugenics) even if it were estimable.

But heritability was the concept that would not go away. It is just so simple to calculate—Goldberger's statistical critique aside. Undeterred, the discipline of behavioral genetics—a subbranch of psychology— has continued analyzing twins, adoptees, and other relatives to estimate the genetic portion of an ever wider range of traits, from bread preferences to socioeconomic status. As we mentioned in chapter 1, the heritability debate boiled over in the 1990s with the publication of Richard Herrnstein's and Charles Murray's bestselling book, *The Bell Curve*. They took the work of the behavioral geneticists in which Herrnstein had been steeped as a psychologist and merged it with the social history and policy analysis in which Murray was conversant, given his role as the leading right-wing pundit on social welfare programs. The result was a very elegant—if scary—argument.

In the bad old days, their story went, the United States (and certainly Europe, too) was a land of unequal opportunity. We were the country of separate and unequal when it came to race, gender, sexual orientation, and so on. Despite the greatest hopes of some of the founding fathers, until very recently we were a society whose privileged elite sent their children to college to become managers while most of the rest of us were lucky to graduate from high school. The Jeffersonian ideal of a nation of smallhold farmers never truly came to pass, and class divisions (along with race and gender divisions) became entrenched.

During the mid-twentieth century, things started to change. Of course, there was the New Deal, which stitched together a safety net so that those at the bottom did not fall too far. And then there were the policies of the postwar period that led to a massive expansion in college enrollment (such as the GI Bill and the establishment of Pell Grants during the War on Poverty) as well as a rise in home ownership (thanks, in part, to Federal Housing Administration and Veterans

Administration loans, the establishment of Fannie Mae, and so on). Meanwhile, compulsory schooling laws were introduced across the country, making high school graduation the norm and not the exception. College access was expanded through an increase in the number of seats and also through a new emphasis on standardized testing for admission, which acted as a counterweight to the old boys' network. Among those attending college, the brightest students now went to the best colleges. Meanwhile, the civil rights triumphs of the 1960s ended the last vestiges of de jure apartheid, and voilà, the Shangri-La of meritocracy was born. At least, this is what Herrnstein and Murray argued.

The ironic twist in their tale is that thanks to the new reality of equal opportunity, those meritocratically selected Yale graduates were now reproducing with fellow Ivy League alumni to produce even more intellectually gifted offspring. Over time, through this process, which demographers call assortative mating, the gap between the genetic potential of those at the top and those at the bottom was widening. In the bad old days, Mr. Boss was said to typically marry the beautiful secretary and not his fellow executive, and thus one-half of the mating equation was being sorted on ability and the other half was being sorted by attractiveness—cue the *Mad Men* theme song here.[6] (We will address this piece of the puzzle in chapter 4, on genetics and marriage.) Throughout the 1960s, Ivy League marriages were on the rise, while the economy was becoming increasingly knowledge based, which favors these folks, and in just a couple of generations, the result was intractable inequality: intractable because it resulted from what Herrnstein and Murray called *genetic stratification*, in which socioeconomic class is the product of genetics rather than societal processes and is thus immune to policy fixes. Their argument culminates to suggest that instead of trying to develop policies to promote equal life chances for children across social backgrounds—which was an increasingly futile endeavor due to their genetic roots—society might as well just give up and pursue a public policy meant to merely avoid unrest among the increasingly genetically disadvantaged at the bottom.[7]

While it is tempting to dismiss this argument as the polemic of provocateurs who were more interested in selling books than contributing to careful scholarship on inequality in modern societies, their argument was not entirely pulled out of the ether. Important scholarship was emerging within the ranks of social scientists that suggested that the estimates of environmental influences may have been wildly exaggerated by scholars up to that point. Specifically, what would ultimately be called the causal revolution in social science was just gaining momentum, and this revolution overturned the dogma of the 1970s and 1980s.

Prior to the economics-led causal revolution, social scientists who were interested in predicting, say, how much education or income a given individual would obtain in adulthood might compare individuals who grew up in families with various income levels. If the children who came from families with an annual income of $10,000 earned, on average, $20,000 when they reached adulthood, but the children who came from families who earned $20,000 a year typically netted $30,000 a year, then we might conclude that raising parental income a dollar per year would yield a rise in offspring income of a dollar a year on average. We could then use that calculation in a cost-benefit analysis of the long-term impact of transfer payments (i.e., welfare).

Of course, it could be (and is) the case that higher-income parents tend to be more educated, older, and married compared with their lower-income counterparts. So researchers would measure these other competing factors and control for them statistically in order to focus attention on the effects of family income separately from other family characteristics. After using the magic of multiple regression analysis to compare offspring from parents who are similar based on age, education level, and marital status, they might find that the (statistically adjusted) effect yields 50 offspring cents on the parental dollar. But the problem that the causal revolution identified is that we could *never* measure all the possible differences between high- and low-income parents and families. There could always be some lurking, third (or Nth) variable that might account for the apparent statistical relationship between parents' income and children's income. Religion? Cultural attitudes toward the acquisition of money? Quality of schooling as opposed to quantity? The list is endless, and to make matters worse, even if we could identify every previously unobserved variable, we could never measure them perfectly.

The bias produced by such measurement error and unobserved heterogeneity (the technical term for these lurking variables), could mean that the effect of parental income is either overstated or understated. But most of the plausible candidates for lurking variables worked in the direction that would lead to the "naive" estimate of the effect of parental income being overstated. The original estimates suggesting that $1 of parental income generates $1 of offspring income might instead generate $0.50 of offspring income, or even no offspring income. And if parental income has no lasting impact on children's outcomes, why have welfare?

This possibility was best illustrated by the sociologist Susan Mayer in her book *What Money Can't Buy: Family Income and Children's Life Chances*, which challenged the assumptions of the entire poverty research community. In this clever volume, Mayer deployed a number of research designs (called counterfactuals and natural experiments) to show that the traditional estimates of the effect of income on children's life chances have been grossly overstated. For example, she showed that a dollar from a transfer payment had little to no effect on children, whereas a dollar from earnings had a much bigger effect—suggesting that it was the underlying attributes of the parents that led them to earn money that were having the positive effect, not the dollars per se. She also showed that additional income did not usually result in the purchase of goods or services, such as books or tutors or health care, that we would expect to improve the human capital or life chances of children. Rather, most unexpected, extra money was spent on consumption for the parents (including vice purchases like alcohol and cigarettes). Although there were certainly limitations to her work and some questionable assumptions in her models, she upended the world of poverty research, and some of her findings suggested that Herrnstein and Murray—who were advocating that children's outcomes were somehow hardwired from genetic factors that were relatively immune to policies (like welfare)—were not as far outside mainstream social science as we might want to consider them to be.

Mayer's research suggests that something else, something that is transmitted between parents and children, must be generating these overinflated estimates in the so-called naive statistical models.[8] If Mayer is right, we see three plausible explanations. First, it could be cultural transmission of knowledge and practices from parent to child that lead to greater productivity within the economy and thus conventional economic success. These pathways might include knowledge about how to navigate the job market or the school system, crystallized intelligence (i.e., accumulated knowledge), an emphasis on hard work, impulse control, or any number of less measureable attributes that are passed on through the nutritional, health, and cultural environments that affluent parents create for their children. This is bad news in terms of income and welfare policy since income transfers do not generally also transmit key knowledge and practices, but it is good news for equal opportunity if we can uncover the transmitted traditions that "work." This is the impetus behind programs ranging from the Women, Infant, and Children Nutrition Program (WIC; to improve the nutritional environment of poor kids), Medicaid (to augment the health of disadvantaged children), and even Head Start, Sesame Street, and new programs encouraging parents to speak and read more to their babies (all of which seek to improve the cognitive stimulation that young ones receive). It may be that there are a thousand and one cultural practices that need to be augmented and instilled in order to level the environmental playing field.

A second possible dynamic that would explain the overestimation of income effects is that each generation of poor (or rich) kids faces anew environmental obstacles outside the home. Consider the correlation of race and income. If educators and employers discriminate against minorities in each generation, it would be no wonder that income gains for parents do not translate into better earnings for offspring. The lack of effect of parental income need not be race, per se; such discrimination could be based on any number of physical or cultural markers. Of course, this explanation is not exclusive of the first. There could be something that is passed on in the home, such as a way of speaking, that leads to discrimination for both parents and children even though it has no impact on actual productivity.[9] The distinction between the first explanation and the second is the assumption that the cultural attributes passed on in the first instance do, in fact, lead to differences in productivity and thus cannot be waved away as irrational discrimination. (Of course, this can become a bit circular in an economy in which cultural production plays an important role, making up almost 5 percent of gross domestic product [GDP] directly). For example, speaking the lingua franca of the dominant culture—standard English, for instance—does affect productivity because of the self-fulfilling prophecy that results from current power brokers preferring that form of language, or clothes, or whatever.[10] Even if generations of poor folks face discrimination, we can still address this root problem through public policies such as income- and/or race-based affirmative action. If either of the first two explanations are the most salient ones, after a couple of generations of careful application of remedial policies and a vigilant eye against any residual discrimination, policies that address the unequal playing field in the home or outside it should no longer be necessary. For example, fixing *Pygmalion*'s Eliza Doolittle's diction should mean that her children automatically pick up the Queen's English, so policy makers can relax. Likewise, if we stamp out de jure and de facto discrimination, then within a couple of decades, individuals from all backgrounds should occupy all rungs of the economic ladder, and bias or prejudicial treatment should occur infrequently.

The third possibility is the most disheartening to liberal policy makers: that the productivity of poor parents is lower than that of rich parents (hence the former's poverty) and furthermore, this difference is woven into the genes and thus passed on, to a large degree, to their offspring. In the hypothetical world where this is the dominant explanation, it would be no surprise that what we do to aid parents has little second-order impact on their children. That said, Goldberger is right that just because the roots of some outcome such as poverty or myopia is genetic does not mean that we should close up the Royal Commission on the Distribution of Eyeglasses. Indeed, we will need to hand out prescriptions for generations to come as long as we do not pursue a policy of eugenics against those with bad eyesight. Likewise, while we may decide to keep open the Royal Commissions for the Distribution of Income Support Payments, all the support is ephemeral and will need to be undertaken anew in the next generation.

WHAT'S WRONG WITH HERITABILITY ESTIMATES

For the reasons discussed above, we think heritability estimates are essential to our understanding of many aspects of society—everything from how meritocratic our educational system is to how effective interventions to mitigate chronic diseases will be. At the very least, such estimates tell us what the multigenerational or dynastic effects of policies will be. That said, if we want to obtain the "correct" figures for the heritability of socioeconomic status, it is essential to understand the approaches that have been used to make claims that, for example, 50 percent of the variation in income is genetic in origin. Is there is a method to the madness or, rather, a madness to the method?

For the longest time—stretching all the way back to the days of the statistician and biologist Ronald Fisher and Galton—twin studies were seen as the gold standard for assessing the genetic component of physical or social traits. Being 100 percent genetically identical, ironically, makes twins a tool for understanding environmental conditions. By holding genetics constant, identical twins help us identify the effects of factors beyond those of genes: One twin suffers from nutritional deprivation in utero thanks to the position she occupied with respect to the placenta (i.e., her sister hogged all the nutrients). One wins the lottery. Another gets drafted. There is a robust tradition of twin-difference research in economics and other related fields.

But to deploy twins to tell us about the genetic component of a trait—like height—we need to use both identicals and their less glamorous cousins, same-sex fraternal twins, to estimate what researchers call an ACE model, in which a trait is broken down into three components. A stands for additive heritability (i.e., the genetic component); C stands for common (or shared) environment (i.e., what exposures siblings share, like their family home and neighborhood); and *E* stands for unique, or non-shared, environment (like the different teachers they get in school or friends they encounter). Variation in a Trait = A + C + E. The method has been around for decades and is fairly straightforward in terms of its algebra, but as we will see, it rests on some critical assumptions that until recently were largely untestable.

The basic idea is that while identical twins share 100 percent of their genes, fraternal twins share, on average, only half of them. Thus, the greater degree to which identical twins resemble each other than do fraternal twins is an indication of how much genetics matters and is known as the heritability of that trait (or, specifically, the additive heritability).[11] As a bonus, we can also parse the environmental contribution to traits into portions that are unique to individuals and shared by siblings.[12] We get the unique portion (E) by merely asking how far away from 100 percent the similarity in identical twins on the given trait or phenotype is. By unique environment (E), we mean the inspiring math teacher one sibling was assigned to in eighth grade while the other had to suffer through an instructor who made algebra as fun as waterboarding. If parents happen to treat siblings differently (to the extent that difference is not induced by their genotype in any way), this is part of E. By shared, or common, environment (C), we mean those environmental factors that are shared by siblings and are not caused by their genotypes; for example, if music lessons are induced by evident, genetically based talent, then they are part of the overall genetic effect (A), but if they are just imposed on all offspring in the family by the parents irrespective of talent, then they are part of C.

By way of example calculations, if identical twins have a 95 percent concordance (i.e., a correlation of .95) in height, then their non-shared environment, E (i.e., the individual-unique component), is 5 percent. If we next figure out the genetic component (i.e., additive heritability), we can then subtract both of these numbers to get the environmental component that is common to sibships (C). So, to continue with our example, if the correlation or similarity for same-sex fraternal twins in height is .55, then the difference between identical and fraternal correlations (.95 – .55) is 40 percent; doubling that gives us an estimate of 80 percent additive heritability. A genetic component of this size, in turn, means that the environmental component is 20 percent, and we already know that the unique, or non-shared, environmental component is 5 percent. Thus, the shared, or common, environmental component is 15 percent.[13] These figures are actually not that far from the consensus estimates for height in the developed world. For our purposes of understanding the biological roots of social behavior, let us focus on the genetic component—80 percent in this example.

The most critical assumptions in this exercise are the following. First, the entire endeavor assumes that we can accurately tell who is fraternal and who is identical. Second, the ACE model rests on the assumption that the fraternal twins do indeed share, on average, 50 percent of their genes—or rather the relevant genes (we will not question the 100 percent figure for identical twins, though some new research suggests that perhaps we should).[14] Third, for ACE to work, we must assume that the environmental similarity for the two types of twins is the same—a postulate known as the equal environments assumption.

For the moment, we will take as given that scientists can accurately distinguish the zygosity of twins (i.e., which are fraternal and which are identical), but later we will return to this point. The second assumption of 50 percent average genetic relatedness of fraternal twins seems pretty solid at first glance. Parents and offspring share exactly 50 percent of their genes since children get one hybrid chromosome from each parent.[15] Nonidentical twin siblings each share half their genome with each parent but not necessarily the "same" half. That is, on average, siblings share half their genes, but some pairs of brothers and sisters can be more or less related than others thanks to two factors. The first factor is luck, since siblings each get half their deck of cards from each parent, but the extent to which the cards in those half-decks overlap with each other is a matter of chance. Some sibling pairs will actually be closer to half-siblings in terms of genetic similarity and others will be closer to the identical-twin end of the spectrum. This is not a problem for the ACE model, since all we are assuming is that, on *average*, fraternal twins share 50 percent of their genome. But a second factor—assortative mating—does affect sibling relatedness in a way that skews results since it moves the 50 percent average up or down. That is, if parents tend to be more genetically alike than they are similar to random other individuals in the population—that is, there is not random mating but rather assortative mating—then siblings (including fraternal twins) are more alike, on average, than the 50 percent we assume. If parents are less like each other genetically than they are to the random stranger, then there is negative genetic assortative mating. We will discuss genetic assortative mating in chapter 4, but the upshot for the ACE model is that if positive assortative mating is the norm, then we wind up with an *underestimate* of the genetic component of a given phenotype. If there is negative assortative mating, our heritability is biased upward. All available evidence points to positive assortative mating both overall and with respect to trait-specific genetic signatures. What this violation of the ACE assumptions means, however, is that heritability is probably underestimated rather than overestimated! Namely, typical calculations assume that the shared genetic component for monozygotic (identical) twins (100 percent) is twice as big as that for dizygotic (fraternal) twins (50 percent). Thus, when there is positive assortative mating, the actual difference is smaller since the genetic similarity for fraternal twins is higher than 50 percent when they are sired by parents who are themselves somewhat genetically similar. Imagine a case in which there is such a high degree of assortment that, on average, the parents' genes for height are strongly correlated with each other—that is, women and men who are genetically predisposed to be tall are more likely to marry each other, and ditto for those who are genetically likely to be short. In this scenario, fraternal twins would be, say, 67 percent genetically similar. Given that the difference between the identical and fraternal twins in terms of genetic similarity is only 33 percent (100 − 67), we should really triple rather than merely double the difference between monozygotic and dizygotic twin similarity to get to 100 percent of the genetic effect. Thus, positive assortative mating leads to an underestimate of additive heritability if we still only double the difference.

The third key assumption of the ACE model—that the environmental similarity for identical and fraternal twins is the same—is an entirely different story. This assumption asks us to believe that for a particular trait of interest, the greater similarity we observe for identical twins vis-à-vis fraternal twins is entirely due to their greater degree of genetic relatedness and not to greater environmental similarities relevant for the trait. What does this mean in practice? Some observers falsely assume that *any* environmental similarity shared by identical twins that is not shared by fraternal twins presents a problem. But it does not. If two identical twins have more similarity in their math curricula because they are equally smart—compared with two fraternal twins who differ on innate math ability and thus are assigned to different tracks—then this increased environmental similarity is the direct result of increased genetic similarity and *should* be captured in the A term. The same is true for experiencing more similar friends because their personalities may be the same, and so on.

The problem arises if people treat identical twins more similarly because clones per se evince a "special" reaction among other people. Imagine a world in which siblings range in genetic similarity between 0 and 100 percent, and in which it is totally normal to see siblings who are 90 percent identical genetically. There would be nothing particularly weird about siblings who happen to be 100 percent related (identical twins). We might be able to assume that the environmental similarity of going from 43 percent related to 48 percent related is the same as going from 95 to 100 percent related. There would be nothing magical about achieving 100 percent genetic similarity (identical twinness) in this context, and the increased environmental similarity would be a legitimate result of genotype resemblance.

However, the world we live in has sharper thresholds of relatedness.[16] Thus, being 100 percent related causes some issues of its own. Specifically, one problem is that identical twins are treated specially: they experience more environmental similarity than is warranted from their genetic similarity. For example, identical twins are more often confused for each other; they also spend disproportionately more time together (each of which violates the ACE model). Such phenomena are qualitatively different from instances in which identical twins are assigned more often to the same math track or have the same number of friends because the twins are more alike innately—which is acceptable within the model. Another model-violating problem with using identical twins lies in epistasis. Epistasis—which we discuss in more detail in chapter 7—is the interaction of different genes with each other to produce specific outcomes. That is, if variation all across the genome interacts to produce outcomes, then only identical twins get the extra similarity of having the exact same multiplicative *combinations* of genetic variants over and above the additive effects of those variants.

These aspects of monozygotic-twin specialness—epistasis, confusability, and greater cross-socialization—mean that there is an effect of being a monozygotic twin that may trump—or at least moderate—the impact of being genetically identical. In social science jargon, identical twins are said to present a problem of external validity—that is, we may not be able to generalize to the rest of the population anything we infer from identical twins, given their special status.[17]

So it would seem, then, that the twin-based estimates of heritability of everything from liking hybrid cars and mustard (37 and 22 percent, respectively[18]) to preferences for jazz and opera (42 and 39 percent, respectively[19]), to education (40 percent[20]), to smoking initiation (44 percent[21]) are flawed. To partially address this problem, researchers tweaked the basic design to produce the extended twin model, in which other relations are included to map out the different levels of relatedness. For example, cousins who share identical twins as parents are—on average—as related as half-siblings (25 percent as opposed to the typical 12.5 percent of first cousins), despite living in separate household environments. If the heritability estimates of comparing twin-offspring first cousins to regular first cousins were similar to those of the standard identical- versus fraternal-twin comparison, then we might have more confidence in them. Of course, these methods rely on similar assumptions about twins and environment—even if one degree removed.

Given the fact that identical twins have experiences that potentially make them nongeneralizable, we took advantage of molecular markers to challenge the twin models by returning to that first, bedrock, assumption: that scientists (and twins themselves) can accurately tell whether they are identical or fraternal. Unlike in the days of yore, when behavioral geneticists had only their own eyes, a few survey questions, and the self-report of families to decide whether a given pair of same-sex twins was identical or fraternal, today we can know for certain by picking highly variable (i.e., polymorphic) stretches of DNA and cross-checking the siblings. (Some stretches of DNA are pretty much the same for all humans but others show large amounts of variation and so are good for solving crimes, doing paternity tests, and making sure that identical twins are really identical.) If siblings match on a set of determined markers (e.g., both have a C or a T at that location or the same number of repeats of a sequence of letters, say, AGG AGG AGG AGG), the chances of that happening by chance are minute. Thus, the twins are probably identical. But if they vary at even one site, they have to be fraternal. Rather than relying on a series of questions that include "whether they were more like two peas from the same pod or, rather, akin to peas from different pods on the same plant" (this is an actual way of classifying twins), we can use molecular genetics to assess zygosity with certainty.[22]

It happens that a significant portion of the same-sex twin pairs we studied were wrong about their zygosity status! In the twin sample of the National Longitudinal Survey of Adolescent to Adult Health (Add Health), 18 percent of same-sex twins thought they were identical when they were really fraternal (or vice versa). The majority of these errors were individuals who thought they were fraternal but were really identical. This is understandable since families may assume that twins need to be completely identical looking to be judged as monozygotic. Yet there is a nontrivial amount of physical variability within identical twin pairs—starting at birth, when birth weights can display dramatic differences depending on the placental architecture and which twin benefited from more nutrients in utero.

If you are a twin, finding out you have been wrong about the person with whom you shared a bedroom your entire life may be traumatic, but for us scientists, the situation provides the ideal test of the robustness of the ACE model to violations of the equal environment assumptions. By using the monozygotics who grew up identifying and being treated as fraternals, we could see whether heritability estimates changed when this environmental "specialness" was absent. The fraternals who thought they were identical provide another cross-check.

Much to our surprise, there was not a single outcome for which using the misclassified identicals in lieu of the correctly classified ones (or the incorrectly perceived fraternals in lieu of the correctly identified ones) drove down the heritability estimates. We did this with three different data sets—two from the United States and one from Sweden. We did it with self-perceived zygosity and researcher-assigned zygosity (by sight and by that series of questions, which tended to be more accurate, with an approximate 5 percent error rate). The results held across all of these methods and samples. We, the social scientists who questioned the veracity of the equal environments assumption and assumed that the behavioral geneticists were making a fundamental error, ended up confirming their "naive" ACE models.[23]

Before we tried, and failed, to knock down the twin method, a set of new methods had already emerged to circumvent the inherent "oddity" of twins. These approaches were made possible by the genomic revolution and the widespread availability of cheap genotyping chips. Now that scientists can genotype large samples of subjects, we do not need to guess the genetic similarity between pairs of individuals (such as siblings) based on a presumed family tree; we can instead actually measure it using these approaches. The strategy involves looking across the entire set of 46 chromosomes and asking how well a pair of individuals match up across the 1 million or so bases that the chip assays. This is how it works: Imagine that there is a sequence of 10 bases instead of 1 million. Each position on the chosen strand (which is arbitrary) has two possibilities (we use only biallelic loci for this example—that is, areas of the genome where the options are only two of the four bases, in this case, an A or a C). We will call one of these bases the minor base (typically the one found less often in the population) and the other the reference base. The point is that we can then count for each individual at each position whether they have zero, one, or two of the minor alleles at that position on the two chromosomes. So at position 1 of 10 positions in our hypothetical example, 49 percent of people in our sample may have an A there while 51 percent have a C. We call the C the reference base, and we go through our entire sample of genotyped individuals and give them a score for the first of the 10 positions on each chromosome: 0 if that person's genotype is CC; 1 if it is AC or CA; and 2 if it is AA. We do that scoring procedure for each of the 10 positions we have measured (of course, the actual bases we are looking for may be different at each position). We can then calculate how similar any two individuals are at each position (i.e., the correlation) and then add up those 10 scores for an overall "relatedness" measure.[24]

If we do this for all the pairs in our sample,[25] we can then ask: do pairs that are more genetically similar also happen to be systematically more similar for our phenotypes of interest? That is, does genetic similarity predict trait similarity in, for example, height or educational attainment? These new methods are not much different from the twin method: compare pairs of random people who are more or less genetically similar and see how much of their phenotypic similarity it accounts for. The only difference is that for the ACE models the genetic differences are assumed to be fixed at 100 percent versus 50 percent. In the new approach using unrelated pairs, we are talking about the difference between, for example, one-half of 1 percent and 1 percent genetic similarity. The benefit of the new methods is that we do not have to make any assumptions about twins being generalizable to the rest of the population—in fact, we are directly analyzing the "rest of the population." One pair may be positively correlated at .01 across their entire genome (i.e., 1 percent more similar genetically to each other than they would be to randomly selected other individuals), another pair may be right at 0, and another may be slightly negatively correlated. And it turns out that published studies using this approach tend to report heritability estimates that are about one-half that of the twin methods. (See chapter 3 for a discussion of the issue of missing heritability.[26])

Once again, we were skeptical about the assumptions underlying this research design. What if the real cause of the correlation between genotype similarity and outcome similarity was not due to the causal effects of genetic variation but to the fact that genetic variation was capturing environmental differences as well? This is known as the problem of population structure, or population stratification. Another term for it is the chopsticks problem, so dubbed by the population geneticists Dean Hamer and Lev Sirota.[27] Imagine a sample of Americans. In this sample you find that a certain marker on chromosome 16 pops out as extremely predictive of chopstick use in your data. In your excitement, you rush to publish your finding about the chopsticks gene. But then your lab mate suggests that you might want to run your analysis separately by racial group. When you do that, you find out that the distribution of that particular allele or marker (see chapter 1) looks like this: whites, 98 percent prevalence of C, 2 percent prevalence of A; blacks, 90 percent prevalence of C, 10 percent prevalence of A; Latinos, 94 percent C, 6 percent A; and here is the kicker, Asians, 18 percent C, 82 percent A. Now, it could be that this gene that you discovered somehow has evolved in East Asian populations and causes them to prefer chopsticks as a utensil to forks. But to test whether this is truly a genetic cause and not simply a historical accident intersecting with allele frequency differences by continent, the relationship should hold up within ethnic groups (even, really, within families—i.e., comparing siblings). Lo and behold, when we run the analysis for whites only, or Asians or blacks or Latinos only, we find the presence of a C or an A predicts nothing about chopstick facility. You would have been a victim of population stratification.[28] You thought you were discovering something fundamentally biologically determined with respect to cultural practice when you really just found a very good ancestrally informative marker.

In brief, while we detected in this analysis previously unaccounted-for population stratification that might bias the results, when we took it into consideration, the heritability estimates hardly changed.[29] (Those interested should read appendix 2.) We were foiled yet again in our social scientific attempts to prove the geneticists wrong![30]

THE SECOND COMING OF HERITABILITY: WHY IT MAY MATTER FOR POLICY

Returning to our account of what may be generating overestimates of the effect of income on children as per Susan Mayer's analysis, it would seem that knowledge of which type of inheritance is at play matters to how we design policy—namely, whether it is environmental or genetic influences across generations that are explaining most of the naive estimate of the effect of income or poverty. Even if we do not know which genes matter for which outcomes, simply knowing that, for example, the likelihood of going deaf in old age is 90 percent heritable changes the policy landscape compared with a situation in which that outcome is 10 percent heritable. Any environmental interventions that prevent or fix deafness in one generation are unlikely to yield any dynastic payoffs in the next generation since the risk inherent in the germ line has not been altered. Likewise, if old-age deafness is largely genetic in nature, we will probably not be able to prevent hearing loss absent genetic screening or engineering but instead will need to keep the Royal Commission for the Distribution of Hearing Aids open for perpetuity. However, if the etiology of deafness is largely environmental—for example, due to loud noise—then we are left with two possibilities akin to those that we discussed with respect to income. It could be that the noise comes from random, external sources, in which case we can distribute ear plugs or institute zoning laws to reduce aural pollution, but we will have to keep applying those solutions for each generation if we wanted the beneficial effects to persist.

Consider the possibility that rather than random exposure to noise, it is the familial environment that is the relevant site of environmental exposure and risk; it is parents who structure the risk environment for children. Imagine that some households are loud, and that once a given generation—for whatever reason—starts to lose its hearing, the household members turn up the volume on the television, start shouting, and generally create an exceptionally noisy environment for the children living there. If early childhood is the critical developmental window for sound exposure leading to long-term hearing loss, then when those children grow up and lose their hearing, they turn up the volume for their own children. There is nothing genetic in this saga, just a cycle of environmental exposure across generations. The good news is that if we intervened with the parents (or grandparents) living in the home of a given cohort of children, we could break the dynastic transmission of hearing loss. We might simply treat the symptoms of the elder generations by giving them free hearing aids so that they no longer had to shout to hear each other. Or, if habits were already too deeply engrained, we might add some behavior counseling to the mix to train them to speak more softly, and so on. The key is that whatever investments we make in the older generation would realize a large return in the filial generation since they would be suffering from less hearing loss when they themselves became adults. Now substitute "poverty" for "hearing loss." Long-term cost-benefit rationale motivates lots of antipoverty programs. The case for these programs is strengthened by evidence that the earlier interventions are targeted in the life course—even and especially prenatally[31]—the greater the long-run savings. But if an outcome such as suicidality, depression, or criminality is largely heritable, these investments may be for naught (or at least they will need to be reinvested in each generation).

Heritability is also, paradoxically, an important measure of fairness. In fact, some scholars have argued that we should aim for a heritability of 100 percent for important socioeconomic outcomes![32] They are not arguing for a dystopian world in which our fates are set at conception (even if this is a necessary corollary), but rather a world in which there is zero effect of environmental influences on opportunity. To them, this dystopia is a fair world. They have a point: Why would we want the family environment to matter for success? For that matter, why would we want the random, nonfamily environment to have any impact? Nongenetic family influences—say an aristocratic title or a monetary windfall from a rich uncle—are exactly what inequality of opportunity is made of. Or if environmental influences beyond the family—for example, getting drafted into the military or being spared that duty—shape our health, economic, and family fortunes to a large degree, that suggests unfairness in society as well, merely one of a different sort.

While we are somewhat sympathetic to the argument that we should seek policies to increase the heritability of economic and social position, we think high heritability is only a necessary—but not sufficient—component of a utopian society of equal opportunity. (And, of course, some may want a socialist society with equal outcomes rather than equal opportunity, in which the influence of genetics and environments is zero; still others may see maximizing E, random, non-shared environments, as the most fair.) How and why genotype matters is also critical to the characterization of whether a society is fair or not. We could achieve a 100 percent heritability of socioeconomic status if we assigned jobs and other opportunities based on skin tone, hair type, or height—all of which are highly heritable. But would that constitute meritocracy?

Or to take another example, if all elite college admissions spots were determined by skill at shooting free throws on a basketball court, and, in turn, this skill turned out to be like eye color—almost 100 percent genetic—then we *could* conclude that such a system of higher education functions as a completely equal opportunity. However, this strikes us (and we imagine you) as illegitimate since almost nobody outside the NBA would suggest this as a rational way to allocate *all* college placements. There is a mismatch between the "test" or "skill" and the nature of the institution. Likewise if the NBA offered starting positions based *only* on performance on the SATs, not many people would want to watch the Lakers or the Knicks.

Luckily, we live in a capitalist society in which the market serves as a disciplinary force that encourages firms to align rewards with marginal value added. So if height were the heritable criterion for running some large corporations, those corporations would probably fail compared with companies that deployed a market-driven policy to promote their executives. The problem is that in a society in which—as mentioned above—4.2 percent of GDP is composed of cultural products whose valuation is arguably not inherent but subject to tastemakers, gatekeepers, and the consumers with the most market power (i.e., the rich), the whole idea of means and ends in economic life can get a tad circular.

Along these lines, some recent research has shown heritability of IQ to vary by such policy-salient categories as race, income, and parental education. For example, Guang Guo and Elizabeth Stearns show that for blacks, the heritability of IQ is lower than for whites.[33] They interpret this to mean that environmental conditions—such as a lack of parental resources, poor schooling conditions, or simply racism—prevent the full realization of genetic variation in this population—

that E is the bigger factor that squelches A in the ACE model. In other words, there is an implied gene-environment interplay effect such that potential intellectual ability is inherited but requires environmental conditions of human capital investment to be realized in the form of IQ.[34] If it is true that blacks demonstrate a lower heritability of IQ than their white counterparts, then here is an instance in which we might want to know heritabilities for various groups as a diagnostic of how and where environmental conditions may be hindering a more efficient distribution of human capital.[35]

In other words, at face value, the lower heritability estimates for African Americans tells us that there is something worth exploring, if not the specifics of why the estimates are what they are.[36] If we discover an interesting empirical relationship between unmeasured DNA and a given outcome of interest to social scientists, let us try to understand what accounts for it rather than assert its irrelevance for policy. In the process of investigating, we are likely to gain knowledge that is of policy relevance. By using these differences in heritability as a starting point, we can then explore the underlying processes by which they arise. Taking the case of race and IQ, each of the above-mentioned explanations for the lower heritability for blacks (racism, resources, and so on) is worthy of investigation, both experimentally—by manipulating the environment through, say, family or school policy to see if heritability estimates change—and through the interrogation of molecular genetic data to determine specific loci that may be responsible for the intergenerational genetic association. Even without identifying the loci that contribute to a given trait's heritability, the knowledge of the conditions under which genetics matters to a greater or lesser extent is important to policy makers to decide whether they want to try to drive that parameter up or down. We discuss this issue more in chapter 7, which concerns gene-environment interactions.

With all these caveats in mind, we hope that you see that heritability is a concept that should not be summarily shunned by scholars. It is a tool—among many—to understand the landscape of opportunity and inequality in a society. In fact, whether we strive for 100 percent heritability or 0 percent, or whether we invest in treating only nonheritable conditions and neglect heritable ones on the grounds that the returns to our investments fizzle after one generation, heritability reveals much about how our society reproduces itself and changes across generations.

LEARNING TO LIVE WITH HERITABILITY

So if many individually flawed heritability estimates converge on a consensus when it comes to cognitive and socioeconomic outcomes (such as education or income), and those heritabilities tend to be around 50 percent or higher, does this mean that Richard Herrnstein and Charles Murray were actually prescient a full decade before the genomics revolution? Not quite. If the heritability of economic status approached 100 percent, and we had faith in the free market to ensure that the criteria for success and efficiency were themselves fair and logical, the world would not necessarily descend into more and more rigid genetic castes. This is because another key element in the *Bell Curve* is assortative mating. As long as potential parents did not sort on genetic potential—that is, the genetically most endowed in terms of IQ and earning potential mating with each other and the genetically least endowed pairing up—then each generation would be reshuffled and genetically based skill would determine position but would not calcify into more static and unequal strata.

Much recent work, however, has documented an increase in assortative mating along socioeconomic dimensions, suggesting that the spread of phenotypic and genotypic variation may be widening as well. (When people mate randomly, their offspring tend to cluster in a nice bell shape around the population average. But when they sort, that bell curve gets flattened out as the ends pull away from each other.) Indeed, if there were a single point to be made about the changing nature of marriage over the past fifty years, it would be that spouses have been picking spouses with increasingly similar educational levels. We will turn to the genetics of marriage in chapter 4, but first we describe a controversy that still rages in genetics: if heritability is so high, why can't we find it in specific genes that matter?

CHAPTER 3

IF HERITABILITY IS SO HIGH, WHY CAN'T WE FIND IT?

When the age of molecular genetics dawned in the 1980s and 1990s, scientists interested in the biological bases of human behavior were thrilled because they could finally open the black box of the genome and measure the effects of genes directly. They no longer needed to rely on often-ridiculed or misunderstood assumptions about twins, adoptees, and the like. Now it was possible to investigate exactly which genes mattered for which social outcomes; they could delve into the biological mechanisms and come up with a better understanding of the pathways from cells to society. They might even be able to develop gene therapies for anxiety disorders, for depression and schizophrenia, and even for poor cognitive functioning. Knowing which genes contributed to the 50–75 percent of variation in these outcomes that they "knew" to be genetic would be the first step in a real clinical agenda for social life. But it turns out that work on the molecular basis of human behavior (and work on most outcomes, for that matter) has endured as bumpy a path as the estimation of heritability did.

In retrospect, it was silly of scientists to think that they would find *the* gene, or even the handful of genes, for sexuality or for IQ. Social life is infinitely more complicated than eye color (which is influenced by three genes). Even something as biologically determined as height (which is 80–90 percent heritable) is highly polygenic—that is, influenced by thousands of genes, each with tiny effects. And if height is affected by that many genes, then social behavior must involve almost every gene known to humankind.

STRIKE ONE: CANDIDATE GENE STUDIES

Even before the genomics revolution of the last twenty-five years, specific genes had been discovered that—if mutated in certain ways—could have profound effects on humans. These genetic disorders, such as Huntington's, are called Mendelian diseases: a certain gene fails to function in the expected way as a result of a mutation that has been passed on, and a disorder results. When a single-gene disease is recessive (that is, it requires two copies of the dangerous version to manifest symptoms)—as in the case of Huntington's or sickle cell anemia—then some people are carriers and show no adverse effects of the genetic variant. But if a carrier mates with another carrier and an offspring gets two copies of the "defective" gene, then symptoms present.

Even cancer can be thought of in this "OGOD"—one gene, one disease—way. Many cancers arise when a tumor suppressor gene becomes mutated so that it no longer functions to tamp down the cell cycle and the cell starts reproducing wildly. If someone has only one working copy of a particular tumor suppressor gene because the other copy was nonfunctional, a mutation in the one working copy (from exposure to teratogenic environmental influences or simply a random copying error when a cell divides) is enough for the cell to start to multiply out of control. An alternative is that a proto-oncogene, a gene that promotes cell growth, becomes mutated (into an oncogene) in such a way that it becomes more active (rather than less active, in the case of the tumor suppressor gene) and thus stimulates cell growth to accelerate. Note that this scenario is a slight oversimplification because a mutation alone is usually not enough to induce cancer. The body has other defenses to prevent cells from going haywire, but sometimes these are breached as well. For our purposes, the point is that a framework that searches for specific effects of specific genes was the operant paradigm in the genetics research community.

Against this intellectual backdrop of known genetic diseases, early research that took a molecular genetics approach to human behavior looked for single gene explanations. Researchers typically investigated this mystery by measuring genetic variation in two ways. First, scientists looked at particular locations in the human genome for single nucleotide polymorphisms (SNPs, pronounced "snip"), that is, variations in base pairs at specific points along a chromosome that are present in at least 1 percent of a population. The other method was to look at copy number variants (CNVs), which are variations in the number of nucleotide repeats of a given pattern or motif. With CNV, some persons may have a string of TTATTATTA (three repeats of TTA), whereas others may have four or five repeats of TTA. It was not just the reigning medical paradigm that engendered this slow approach to gene discovery. The candidate gene method was also partly driven by the expense of genotyping and partly by good, hypothesis-driven science.

Cost was no small factor: In the early days of biological, medical, and behavioral research on genes, strands of nucleic acids called primers had to be created for the genetic sequence under investigation.[1] It was an expensive process, so researchers used great care in choosing where to look for effects. The expense made scientists provide a very good reason for suspecting that a particular region of the genome was key to understanding variation in a particular phenotype. This was no throw-the-spaghetti-against-the-wall approach to research. (That would come later.)

So how did researchers know where to look for effects on the outcomes about which they cared? For the most part, gene selection was based on research that had already been conducted on model organisms—that is, lab animals. Lab animals solve many of the problems of doing research on something as complicated as the genetics of behavior. First, scientists can custom design their environments. One mouse can be assigned to a stressful condition (e.g., taken from its mother before it is weaned) and another mouse can be assigned to the control condition (left with its mother). Random assignment to environments (or just keeping the environments the same for all lab subjects) mimics the gold standard of the randomized control trial in medicine—it removes those worries we discussed in chapter 2 about genes capturing differences in environment; for example, that Chinese Americans use chopsticks and have an overrepresentation of C at a certain location. In lab animals, the problem of gene-environment confounding is eliminated through researcher control of the environment.

In addition, even genetic control is possible in many cases of research with model organisms through a process called back-breeding. That is, by breeding animals with their parents (or siblings), scientists can, over the course of several generations, eliminate most genetic variation, thus creating a race of twins or clones in the lab. Within this isogenic (having the same or similar genes) background, researchers can then change one gene via the transfer of genetic material into the host or via targeted mutation using a variety of technologies. Once inserted into the germ line (the cells that produce the sperm or ova), the genetic alteration will be inherited for generations.

This ability to genetically manipulate live animals opened up a plethora of possibilities. Scientists could not only control the environments of their lab rats within the laboratory setting, now they could insert specific genes (or turn them off) and see what happens. They could also fuse novel genes to existing ones to serve as markers of when and where in the animal those genes were expressed (produced proteins). For example, green fluorescent protein (GFP)—which is native to some species of jellyfish—is now ubiquitous in genetics labs as a marker. When the part of this gene that codes for the fluorescent protein is inserted along with the regulatory portion of another gene, it acts as an indicator that the gene targeted for measurement is turned on.[2] This process allows researchers to see in which cells, at what stage of development, and by which environmental factors a given gene is activated.

Given all of the detailed studies that can be done with lab animals, it should be no surprise that when deciding where to look in the human genome, researchers interested in the genetics of behavior initially took their cues from studies of transgenic mice and rats. The good news is that from the point of view of the biosphere, mice and humans are practically like identical twins—if you think about other creatures we *could* compare ourselves with, such as slime molds and the microorganisms on hydrothermal vents at the ocean floor. Mice and humans share a common ancestor from 80 million years ago, the same brain structures, and almost all the same genes (of 4,000 studied, only 10 differ between the species), and 85 percent of their protein-coding DNA sequence is identical.[3]

Even more exciting for those of us interested in behavior is the fact that our little, four-legged cousins have well-studied behavioral phenotypes. There are ways to measure addiction (for instance, when, like a human drug addict, mice ignore food, sex, sleep, and anything else to press the lever for more cocaine); a series of murine behaviors called social defeat,[4] which is supposed to approximate depression in humans; and a behavior that mimics what we call anxiety. Scientists are even able to measure cognitive function and tenacity (aka grit), which has recently been argued to be a key noncognitive skill for humans in today's society.[5]

So, when variation in certain genes in mice seemed to show effects on murine depression levels, human molecular geneticists decided to study variation within those same genes in the human population. Such genes include those that are highly expressed in the brain and are, today, targets of multiple drug therapies. For example, one candidate gene that has been highly studied in both mice and humans is the serotonin transporter, which is also the target of Prozac and other selective serotonin reuptake inhibitors (SSRIs). Others include the dopamine receptors 2 and 4. Dopamine is key to the reward and pleasure circuits of the brain and has been implicated in attention deficit hyperactivity disorder (ADHD) and its therapies (which include amphetamine stimulants to increase dopamine release). In theory at least, human behavioral scientists had many reasons to look where they did to find genetic influences on social outcomes.

Theory is one thing; practice is another. Many pioneering survey studies that collected human DNA—such as Add Health—measured variation in (i.e., genotyped) 6 to 10 known markers that were part of the serotonin and dopamine system (along with monoamine oxidase, which was the target of the pre-Prozac antidepressants). Many early studies (including some conducted by the present authors) found significant effects of such candidate gene variation in the human analogs to mouse behaviors. For example, as a result of these studies, variation in the gene monoamine oxidase A (*MAO-A*) is thought to regulate mood and aggression in humans;[6] it has often been called the "warrior" gene.[7] Both the gene for dopamine receptor 2 (*DRD2*) and the gene for dopamine receptor 4 (*DRD4*) have variants that have been shown to correlate with human behavior. Some individuals, it is thought, need more stimulation to achieve a given level of response in this part of the brain and thus are more prone to take risks.

There are many important caveats in interpolating results from animal studies to human. One obvious problem with the mouse-guided human-research approach lies in analogizing human behavior with animal phenotypes. How sure can we be that a certain genetic variant that makes a mouse curl up in the corner of its cage is equivalent to that same allele that causes a human to score above the clinical threshold for depression on a survey scale? Or that freezing in fright when a cat appears or a sound associated with a shock is played is the same as human anxiety disorders that manifest as obsessive thoughts and sleeplessness? A second issue is that animal studies control not only the environmental conditions—drastically limiting any unwanted noise that might hinder the ability to detect the genotype-phenotype relationship—but also the genetic background by using isogenic animals to eliminate gene-gene interactions, also called epistasis. A gene-gene interaction occurs when variation in one gene has a noticeable effect only if a second gene also has a certain genotype. For example, it may be the case that variation in *DRD2* does not matter unless you also have a certain variant of *DRD4*, because the two receptors may act as substitutes for each other, one gene producing more receptors when the other gene is not producing enough. Finally, as already mentioned, an issue with candidate gene approaches is that it is difficult to eliminate the chopsticks problem, because the single variants being studied tend to appear at different rates across different populations and subpopulations, both of which also have very different cultures and histories.

Despite these challenges, many studies have found significant impacts of candidate genes on outcomes ranging from depression and grade-point average (Conley) to test scores and ADHD (Fletcher).[8] And those are just some of *our own* studies. There are literally thousands of published results purporting to show the importance of this or that gene on one or another human behavior or attitude. We (and others) were not completely naive. We tried to deal with some of the issues raised above. For example, to deal with the chopsticks problem, we generally conducted analysis within ethnic groups or by comparing siblings who had variations in the gene in question, thereby completely ridding the study of any population stratification that may have been present in the data. Meanwhile, we assumed that mouse models of behavior were not perfect analogues, but we thought that, if anything, the poor cross-species translation of phenotypes was working *against* us finding any effects. Namely, if we inferred that a particular mouse behavior was the result of genetic effects and tried to correlate that behavior with certain human behavior, we would fail to discover those same genetic effects in humans because the behaviors were not congruent (i.e., there was a high degree of measurement error).

Despite our initial excitement and our theory-based study design, the consensus today is that most of these early results were false positives—that is, they were statistical quirks that were not true sociobiological effects. If we have all this theory and we use the theory to test a hypothesis that focuses on a single variant's association with an outcome, such as a test score, is that not how good science is done? Do not all the biases in translating from animals to humans work against us finding true positives? Should not false negatives be a bigger concern? One might think that arriving at false positives would be unlikely in practice—somewhat like finding that needle in a haystack by randomly shoving our hand into one area of the straw. Making many attempts that come close to the needle, or even touch it but do not actually feel the pinprick, would seem to be the bigger concern (false negatives).

One explanation for how we ended up with file cabinets full of false-positive results can be understood with a quick dive into how social science is produced, published, and received in the press, where the focus is on novel findings that provide sexy headlines. With decades of social science publications and tens of thousands of investigators all working on the same data sets, it would seem to be difficult to produce a novel finding. But then along come new variables, i.e., genetic markers added to traditional social science surveys, and voilà, a new spate of findings can emerge. These huge data sets give us literally hundreds or thousands of variables (survey questions) to try out and see if there are statistical associations with the new bits of information (genetic markers). Indeed, in new areas of inquiry—as the hunt for genes associated with complex human behaviors was twenty years ago—investigators might think they should be able to quickly discover new and important statistical relationships that reflect true causal relationships in the data. And there were early successes, such as the discovery of the links between the gene for apolipoprotein E (*APOE*) and Alzheimer's disease and *BRCA1/2* and breast cancer, suggesting that many more outcomes would have big single-gene effects.

There is, however, a risky, unintended side effect of incorporating genetic data into large social science studies. Unlike many medical studies, which focus on a specific disease, social science data sets tend to measure thousands of outcomes, such as income trajectories, educational attainment, political participation, test scores, and the like. If researchers then introduce a set of potentially interesting genetic variants that are thought to be important components of broad biological systems, like the dopamine system, and at the same time do not have a very specific theory to guide their investigations,[9] then they can test and retest and retest associations between gene X and outcome Y (remember, there are 1,000-plus outcomes in these data) until they "find" something. And if the researcher does not find something for the full sample of individuals in the data, maybe she will find something in a sample of men or among whites or those who live in the South. But all these intermediate steps of analysis are not typically reported to anyone. Rather, after tens or hundreds of analyses, the researcher may report only the one or two results that showed an "effect" of the gene on an outcome of some interest. We would love to provide concrete examples, but the point is that we cannot. This is because all the "non-findings"—the sausage refuse, so to speak—ends up on the abattoir floor, where it is quickly swept away. Only exciting, positive findings tend to get published. Those studies that find "nothing" end up languishing in desk drawers. In science, this is called the "file drawer problem," or publication bias.[10]

Unless a result is making such a big claim or is so controversial as to invite independent researchers to seek to replicate it, most go unchallenged in the literature. After all, journals and the mainstream media like to highlight splashy findings such as the "gay gene" and not the careful studies that repeat what the original researchers did but fail to achieve the same results. It turns out that it is very difficult to show conclusively that an old result is "wrong," so these much-less-sexy findings are framed as a failure to replicate old results. More important, however, is that what we see in the journals represents a small percentage of the actual statistical tests run. Increasingly, researchers are encouraged, or in some cases, required, to preregister their hypotheses (i.e., which markers they will test) on a public website to prevent this sort of practice.[11]

For all of these reasons, a backlash against candidate gene studies has emerged. It is just too easy to find something that is not robust and replicable. The result has been papers with titles such as "Most reported genetic associations with general intelligence are probably false positives,"[12] which show the general failure of a result found in one data set to be replicated in other data sets. The problem of false positives with candidate gene studies has been revealed to be so great in the field of behavioral genetics that the flagship journal of the field now no longer accepts such studies, even when they are replicated in independent samples!

STRIKE TWO:
GENOME-WIDE ASSOCIATION STUDIES

So now what? Given the nature of the way science is done and the epic failure of early genetic associations with human behavior to withstand scrutiny, should we have just closed up shop? Should we have decided that the effects of genes on complex human phenotypes were just too contingent, too locally determined by environmental context and genetic background to merit investigation?[13] If we wanted to continue to pursue an understanding of the genetic architecture of important social outcomes, how could we do that in a way that would produce lasting, more meaningful results? Fortunately, at the same time as the wholesale rejection of candidate gene studies was brewing, the price of genotyping was dropping dramatically (see figure 3.1). Both forces inspired many (though definitely not all) researchers to abandon the candidate gene approach and instead cast an atheoretical net across the entire genome to see what they could catch. The era of candidate gene analysis was eclipsed by the period of GWAS—genome-wide association studies.

Genome-wide association studies have been made possible by genotyping SNP chips. Rather than measuring one small stretch of DNA informed by experiments in animals, these measure hundreds of thousands or more alleles like scattershot across the genome (most chips today tag over a million SNPs). For what it cost to test for eight candidate genes a decade earlier, the researcher could now look for the effects of a million SNPs on the outcome of interest. Rather than work from animal models, scientists could throw thousands of strands of spaghetti against the wall and see which stuck—that is, conduct a hypothesis-free investigation and see what "pops out" in the data. The chips are designed to include polymorphisms that show common variation in the human population. The bad news is that since we are more or less conducting a million different statistical tests—by looking at each marker in turn to see if it is associated with the outcome of interest—we need a very stringent statistical threshold of significance to avoid false positives. Typically, a result is considered "real" if it has a less than a 1 in 20 probability of occurring by chance. However, by that standard, 50,000 markers of our 1 million should appear to matter—falsely—just by chance. So a much more stringent statistical threshold is needed. That turns out to be 1 in 50 million.

Even with rigorous statistical thresholds, investigators still need to make sense of estimates from hundreds of thousands (if not millions)

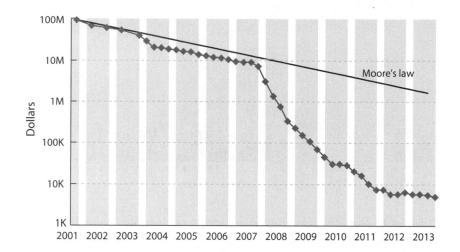

Figure 3.1. The cost of genotyping (full genome sequencing) has been dropping faster than the price of computing power (Moore's law). The price for a gene chip (with about 1 million SNPs) rather than full genome sequencing of all 3 billion base pairs (shown here) is now less than $100. Wetterstrand, KA. DNA Sequencing Costs: Data from the NHGRI Genome Sequencing Program (GSP). Available at: www .genome.gov/sequencingcosts

of statistical analyses. The results from tests of each of the markers are often shown in what is called a Manhattan plot, as depicted in figure 3.2.[14] (Hopefully, your plot looks like Manhattan, with some tall buildings, and not like Paris, with everything flat.) Each shaded dot represents the effect of variation at a given SNP location (e.g., the effect on the outcome of having an A versus a T at base-pair number 12,256 on chromosome 20). The million or so SNPs are lined up by location across the chromosomes, with chromosome 1 to the far left and chromosome 22 to the far right along the horizontal axis. (Some analyses also add the sex chromosomes, X and Y, to the far right of the chart.) The *y*-axis (vertical axis) represents how statistically powerful that particular SNP is in registering a difference in the outcome when the two observed alleles are compared—in other words, how big the

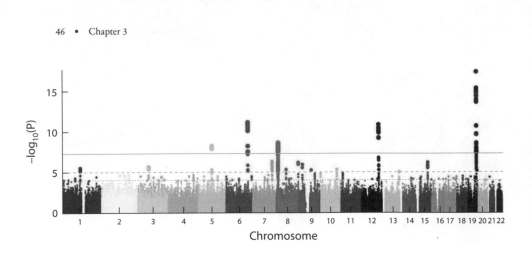

Figure 3.2. Manhattan plot of results from a genome-wide association study (GWAS). Note: The nice thing about the plot is that not only does it show the most statistically powerful SNPs, it shows them in order. So if something is a false-positive, it is likely to be a lone dot, hovering way above all its neighbors because it was likely found in error thanks to chance or a genotyping mistake. However, if it is a "true" result, it is likely to have a bunch of dots in the same area that look like they are climbing up to the peak. These dots are results from independent tests of association for different SNPs with the outcome. Since those that are near the most significant one are also sending strong signals of association, it suggests that that region is truly associated with variation in the outcome. This is due to the phenomenon of linkage disequilibrium (see chapter 5 for a more thorough explanation), whereby SNPs near each other on a chromosome can act as proxies for each other. So the signal gets stronger as we approach the biggest hit and fades as we move away from it. This can be seen clearly in the Pollock-like drips of powerful SNPs in chromosome 19 at the right end of the figure. Visscher, PM, Brown, MA, McCarthy, MI, Yang, J. (2012) Five years of GWAS discovery. *Am J Hum Genet* 90(1): 7–24. Ikram, MK, et al (2010) Four Novel Loci (19q13, 6q24, 12q24, and 5q14) Influence the microcirculation *In Vivo*. PlOS One 6(11): 10.1371

effect is.[15] In figure 3.2, the biggest effect is the uppermost dot on chromosome 19. (The shades are meant only to make it easier to tell the chromosomes apart.)

One advantage of measuring thousands or millions of markers is that researchers can control for population stratification. By factoring out markers that tend to vary together because of shared ancestry, we can be more certain that any residual markers that varied along with the outcome of interest were really causing that phenotype, rather than just reflecting common culture, as in the chopsticks example. In early single, candidate gene studies, there were no other markers to deploy in this manner, but with a million, population structure (i.e., shared ancestry) can be measured and eliminated statistically.

In addition to the ability to tackle population stratification and the more conservative statistical threshold that emerged on the scene with GWAS, replication became more of a norm in this field as well. Once you had discovered the potentially important SNPs that were statistically significant beyond a one-in-a-million chance, you would then perform a second analysis on a "replication" data set with which you would conduct a set of targeted tests—often a dozen or two—in a second (replication) sample that only examines the few variants that looked promising in your first (discovery) data. So, instead of finding 50,000 false results, you might find 1 (or more likely, none). Thus, effects were expected to be demonstrated not just in one, but in at least two separate studies.[16] Thus far, this procedure has produced findings that seem to hold up across time and place and are not sta- tistical artifacts—even if the magnitude of effects in replications tend to be smaller than in the discovery analysis due to the phenomenon of "winner's curse," or regression to the mean.[17]

One surprising consequence of this new methodology, however, was that almost none of the candidate gene results could be replicated or reach genome-wide statistical significance—population stratification and publication bias being the main culprits. So, we had to recon- sider what we thought we knew about the genetics of behavior.[18]

The second big disappointment of the GWAS era was that while some genetic variants did show robust associations with phenotypes of interest, their effects were very small, especially when it came to social and behavioral outcomes. (This result made the original candi- date gene studies, which often showed very large effect sizes, all the more improbable.) When researchers took all the polymorphisms that reached genome-wide significance (less than a one-in-a-million probability of having been seen by chance) and summed their effects (their contribution to explaining variation in the outcome in the population), they arrived at a sum that was nowhere near the calculated additive heritability of traits. For example, the explained portion of heritability for height was 5 percent using the SNPs from the initial GWAS.[19] And with that was born the case of the missing heritability—so intriguing a riddle as to merit a cover of the journal *Nature* in 2008.[20]

As with any good mystery, suspects abounded.[21] One plausible hypothesis was that the SNP chips that were typically used for GWAS covered only the most common genetic variants—by design, for economic efficiency reasons. Maybe, it was theorized, the really big effects that would account for the 90-plus percent of the missing heritability were lurking in this ungenotyped matter. Perhaps once we switched from chip assays to full genome sequencing that assessed each of the 3 billion base pairs, we would find those effects, and the mystery would be solved. Critics pointed out that these rare allele effects would have to cause huge shifts in the outcome of interest, not just because there was a lot of missing heritability to account for, but also because any given allele's contribution to overall explained variance depended on two factors: (1) how strong the effect on the outcome was of that observed genetic change at that particular location and how common it was—and (2) rare variants are, well, rare. If the difference between A and G at a given position in a locus had a small effect but its distribution was 50–50 in the population, that difference would probably have a larger impact on explaining the variation in the population than would the difference between C and T at another location if C (or conversely T) only occurred 0.1 percent of the time even though the effect of C was huge.

This distinction has often led to confusion in interpreting genetic markers. Take the *BRCA1* mutation, which predicts breast cancer, for example. If you have the deleterious allele, your risk of developing breast cancer over your lifetime is eightfold that of a person with no copies of that polymorphism. Clearly this is a risk that someone carrying this gene should pay attention to—as illustrated by Angelina Jolie's bilateral preventative mastectomy (and oophorectomy). However, *BRCA1* explains only a small portion of the overall heritability of breast cancer, since there are many other (genetic) ways to develop breast cancer. Ditto for the *APOE4* allele and its association with Alzheimer's disease. And these are not even terribly rare alleles. It is just that these phenotypes are highly polygenic, which means that they are influenced by many genes. It turns out that polygenicity is the rule and the single-gene disease, like Huntington's disease, is the exception. Even if none of the candidate gene studies had been false positives, it would have taken us thousands of years to add up enough studies to obtain anything close to the actual heritability of social and behavioral outcomes using this hunt-and-peck approach to genotyping. We would have been like the proverbial 1,000 monkeys locked in a room of typewriters for a million years: Eventually we would have written Shakespeare's oeuvre, but it would not have been the most efficient way to produce literature.

Another possible explanation of why we cannot find the effects we are looking for rests in the possibility of nonadditive effects. Typical measures of heritability are called "additive" because they do not take into account nonlinear effects of alleles at a given locus (i.e., dominance). Examples of physical traits that are known to be affected by dominance include brown eyes, dark hair, curly hair, widow's peak, dimples, freckles, unattached earlobes, and double-jointedness.[22] Take the case of sickle cell anemia. The mutation that leads to sickle cell anemia is actually advantageous when only one copy is inherited, since that copy has a protective effect against malaria. However, when an individual has two copies, it is devastating. The sickle cell allele stays in the population in malarious areas due to its advantage in the heterozygous form (when an individual has one copy). This is an example of dominance, or nonlinearity (specifically of heterozygote advantage). The difference in fitness rises in going from zero to one copy of the relevant allele but then declines (severely) from one to two copies.

Single-locus dominance is excluded from additive heritability calculations and so should not be causing the missing heritability problem; however, other forms of nonlinear effects may affect where we set the expectations of heritability. Namely, if the effect of one SNP is contingent upon another SNP, then we have a gene-gene interaction effect. As an example, we can think about the dopamine receptors again. It might be the case that if you have a problematic version of the DRD2 receptor, it has zero effect if you have the functional version of DRD4. If these two genes act as substitutes for each other, then as long as you have one working dopamine receptor, you are okay. Only if you have two "defective" dopamine receptors does an effect on the phenotype of interest emerge. This is epistasis in action.

Transitioning from the hypothetical to the actual, we can think about asthma. Asthma is a disease of bronchial inflammation that is typically mediated by an immune reaction. Interleukins—signaling molecules in the immune system—play an important role in such a response. While certain polymorphisms in the interleukin system confer greater risk of bronchial inflammation, some researchers have found that individuals with a given polymorphism in both the interleukin 13 gene (*IL13*) and in the gene for the receptor to which the product of that gene bonds (*IL4a*) have a multiplicatively greater risk of developing asthma.[23] This sort of gene-gene contingency is not difficult to understand if we think of genes as enmeshed in a robust network that does not look very different from a social network. In fact, 93 percent of genes in the human genome are in some way connected. (And this was what we knew in 2005, so more connections, perhaps linking every gene, may now be known.) Some of these links are known biochemical interactions between the proteins themselves. Others have been shown to be co-expressed—that is, when one gene is upregulated in a cell, the other also tends to go up in activity (or, for that matter, down). In practical terms, this means that if you tweak a gene in one corner of the network, it may have unanticipated consequences in the actions of other genes. For example, if one gene is underexpressed due to a particular mutation, other genes may compensate for its relative absence of protein product. All the indirect pathways here evoke a deeply enmeshed web of paths, suggesting much possible compensation, or many gene-gene interactions. Of course, such a network could also suggest catastrophic effects if key genes were nonfunctional or overexpressed. We can see this in cancer.

With respect to heritability, a genetic network view of human variation (rather than an individual gene-centric approach) suggests that the missing heritability may lurk in such interactions. To return to the two-receptor dopamine example, imagine a case in which having one version of DRD2 has a negative effect on IQ if you have a particular variant of DRD4, and the opposite, positive effect on IQ if you have the other variant for DRD4. In that case, the net effect of DRD2 in a GWAS could be zero. Yet the four combinations may be driving a significant part of IQ heritability. The proliferation of these interactions predicting a phenotype run counter to the assumptions of creating heritability estimates in the first place, and they have been shown to be a potential explanation for missing heritability by creating what geneticists call phantom heritability.[24]

That said, as compelling as the reasons are to suspect gene-gene interactions as a big culprit in the case of the missing heritability, there are also good reasons to dismiss them. If such gene-gene interactions mattered in a huge way, then siblings of two heterozygous parents would probably look a lot less like each other than they do. Revisiting the case of the two dopamine receptor genes, siblings would share allele 1 half of the time, and they would share that second allele half of the time—so one-quarter of the time they would be similar for the combination effect.[25] Extend this logic to three- and four-way genetic contingencies/interactions—and siblings might as well be strangers. But that is not what we observe. We observe a linear progression of similarity as we look across pairs of individuals who are more and more closely related (say moving from cousins to siblings to twins). This suggests that the "additive" heritability estimates we are seeking to match with our molecular investigations are indeed reflecting additivity and not the result of gene-gene interactions in any major way.

Furthermore, there have been good, mathematical demonstrations of why additive variation rather than interactive variation is probably the way evolution works. One intuitive way to think about this is that if every gene depended on ten others, evolution would be pretty constrained and inefficient because any move to a better state would require multiple changes at the same time. Returning to the dopamine receptor example, imagine that having a new version of DRD2 conferred a selective advantage only when DRD4 also had a certain allele, but when DRD4 had the alternate allele, the effect was null.

We would see a better state of affairs for the species only if a given individual had the beneficial mutation in both locations in the genome. And since mutations are random, the chances of that happening in a given individual are slim.

Thus, gene-gene interactions are not likely to be a major factor explaining inherited phenotypic variation in the population. If all effects of given genes were highly contingent upon the state of other genes, we would most likely be locked in place, trapped in a spider-web of genetic networks. Our only hope of evolution would be that those 1,000 monkeys all hit the right keys on their typewriters at the same time. Or, rather, that among those 1,000 monkeys, there was one who had a huge advantage over all others and could carry on the species when the environment drastically changed or who could use the environment in new ways, creating stone tools, using fire, and eventually building the Internet.... Thus, the bottom line is that while some gene-gene interactions may be critically important for particular outcomes (and while gene-gene interactions are important to rule out when trying to demonstrate gene-by-environment interactions), they are unlikely to play a significant role in the heritability story.[26]

Yet another explanation for missing heritability is that maybe Darwin was wrong and Jean-Baptiste Lamarck was right: perhaps environmentally induced changes can, in fact, be inherited. Lamarck was ridiculed for suggesting that the giraffe got its long neck due to efforts to stretch that, with each successive generation, passed on a slightly longer base length to offspring, who continued their efforts to reach the highest leaves.[27] Darwin famously rejected such inheritance of ascribed traits and said that random mutation combined with competition for survival led to phenotypic changes that diversified lifeforms into various ecological niches. However, as of late, Lamarck's idea of inheritance of acquired traits has been making a comeback within the burgeoning field of epigenetics. Specifically, in addition to our DNA code, there is a second, epigenetic, code that allows cells to turn genes off and on in different tissue types, at different times, and in response to different conditions or stimuli. It has long been thought that the epigenome was wiped clean with each new generation—a fresh start of sorts—in order to allow a single cell to grow into an entire human. However, some scientists now believe that epigenetic marks may be inherited. If so, typical molecular measures of genes—which look only at the bases and not the epigenetic marks—would miss this important form of inheritance, thus leading to a missing heritability problem. That said, there is little evidence at the present time that humans can inherit environmentally induced epigenetic marks. Second, even if they could, such marks would not likely generate the high heritabilities in twin models that we observed in chapter 2. In appendix 4, we discuss some of the ongoing findings in epigenetics that have implications for the missing heritability search, but we ultimately come to the conclusion that epigenetics is not the answer.

CHIPPING AWAY AT THE HERITABILITY GAP: THE COUNT IS 0 AND 2, TIME TO BUNT?

In the meantime, there is a much simpler explanation for missing heritability: small effects and small samples. It is similar to seeing the increase in pixels of your pictures when you buy a new iPhone. If we think of the sample size in a genetic study like the number of pixels in a photo, what we recognize is that with a small number of pixels, we can broadly see the "big" features of a picture—whether the picture is of a mountain or an ocean, whether the sky is blue or pink. As we increase the number of pixels (the sample size), we can make more precise distinctions, and we can see minute differences between pictures (this is the move toward a paradigm of "small effects" genetics).

Since most of the outcomes in which social scientists are interested are influenced by many genes, these small effects are rather hard to identify in conventional-sized social science studies (i.e., 1,000 to 10,000 subjects, at most). The photographs are just too blurry to find what you are looking for. Even taking something we are confident is highly heritable like height, we can see in figure 3.3 that by increasing our sample size tenfold—say from 20,000 to 200,000—we also go up an order of magnitude in the number of significant loci found, in this case from just over 10 to just over 100.

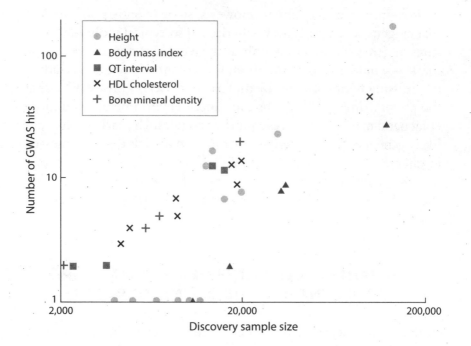

Figure 3.3. Relationship between sample size and number of genome-wide significant SNP effects found in a GWAS.

But the sample-size problem is only half the story here. When we look for specific hits, restricting ourselves to those that reach genome-wide significance levels (5×10^{-8}) is a good idea in order to avoid false positives for all the reasons discussed above. But when looking to explain variation in the population, it is a really bad idea. By deploying such a stringent statistical standard, we are throwing out a lot of useful information (hence, erring on the side of false negatives). This heuristic is akin to the principle in criminal law (sometimes called Blackstone's Formulation) that it is better that ten guilty persons escape than one innocent suffer and be imprisoned. While a reasonable principle, it leaves many crimes unsolved (and genetic effects hidden).

An alternative to throwing away this information is to combine all the information we can glean, and it is simple to do so. We take the

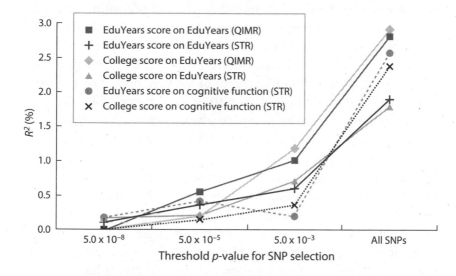

Figure 3.4. The predictive power of a polygenic score for educational attainment, by threshold level of SNP inclusion. Rietveld, CA, et al. (2013) GWAS of 126,559 individuals identifies genetic variants associated with educational attainment. *Science* 340(6139): 1467–1471.

results of our GWAS and add them all together. That is, if for SNP 1 on chromosome 1, we find that each additional reference allele raises IQ by 0.1 points, we then score each person with either 0 if they don't have that allele at the given location, 0.1 if they have one copy, or 0.2 if they have two copies. Then we do the same calculation for SNP 2 on chromosome 1 and so on, through all 23 chromosomes. In this way, we can scan the entire genome and calculate a single polygenic score from our millions of separate GWAS analyses. Then we take the formula for this score that has been calculated from the GWAS of one (or more) sample and calculate it for the subjects of another, independent sample to do the prediction exercise, seeing how well it explains the phenotypic variation (i.e., what additive heritability it yields).

Indeed, as shown in figure 3.4, when restricted to genome-wide significant hits, our polygenic score for educational attainment explains almost none of the variance in either years of schooling or cognitive functioning. However, as we steadily relax this threshold—until eventually, we include all the SNPs in the "discovery" phase of the analysis—we find that the predictive power steadily increases. But still, we explain only about 3 percent of the variation in education, a phenotype that we think is at least 10 times that heritable. (A newer score now explains about 6 percent of the variation in years of schooling, and an effort is underway to add additional samples to further improve its predictive power.)[28]

In the analysis of height, however, researchers can now account for 60 percent of the heritability with a single score that results from the summation of GWAS results—just one number![29] As more and larger samples become available for meta-analysis (a statistical technique that allows pooling of results from multiple samples), this figure is sure to rise.[30] Figure 3.5 shows some projections based on the sample size of a single study (pooling studies makes this slope a bit flatter, but the general idea is the same).[31]

There are still obstacles to reducing the full additive heritability estimate to a single score that can be ascertained by using SNP chips to genotype individuals. One such obstacle is that there is incomplete linkage of the measured SNP markers with the SNPs that really matter in the genome.[32] A second impediment arises from the assumption that heritability varies across time and place. By capturing only those effects that are common across many studies conducted in vastly different contexts with very different cohorts of subjects, we may be capturing only a part of a trait's heritability in a particular sample.

These challenges aside, the basic story is coming into focus: there is no missing heritability. It was hiding in plain sight among the common variants in the human population; we just needed huge samples to be able to amplify the signal-to-noise ratio in our analysis. Those massive samples are coming online now. The personal genomics company 23andme has now genotyped over 1 million individuals— albeit they are unrepresentative, highly self-selective respondents. Kaiser Permanente in California (the health insurer) has more than

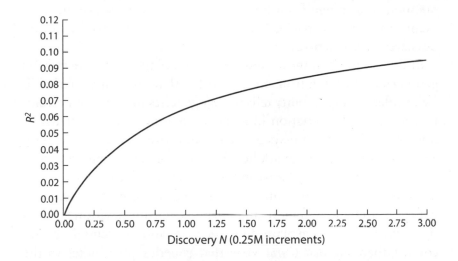

Figure 3.5. Relationship between sample size and proportion of heritability explained by a polygenic score for a continuous outcome for a trait with an SNP-based heritability of 12 percent. Note that the proportion of heritability explained by a PGS conditional on sample size increases with the heritability (h^2_{snp}). The y- or vertical axis represents the proportion variation of the phenotype explained up to the maximum of 12 percent in this case. The x- or horizontal axis measures the sample size required to achieve that prediction as we move from zero to three million genotyped subjects in the discovery sample. Based on data from Daetwyler et al (2008).

500,000 respondents genotyped.[33] The UK Biobank provides genotypes for 500,000 respondents, who were chosen by proper sampling to insure national representativeness—though with a potentially worrisome low participation rate.

When the predictive power of these single polygenic scores starts reaching double digits, a serious policy debate about artificial selection (aka personal eugenics) will be needed, particularly as more and more parents opt for in vitro fertilization. (We discuss policy implications in chapter 8.) In the meantime, by allowing us to directly measure the effects of given genotypes, rather than merely inferring them from relatedness, these scores have become useful for answering

questions that range from how genes interact with social environments (the topic of chapter 7) to how we may or may not be sorting
ourselves into genetic castes.

In this vein, polygenic scores for education allow us to revisit the
genotocracy question from an angle other than the simple heritability calculation. Heritability tells us that genetics matters for a certain
percentage of the variation in a given outcome; it does not tell us
how that genetic effect plays out across and within families. But with
a polygenic score, we can ask how education-related genotypes play
out within families and between families. If the education score predicts strongly between families (i.e., among randomly selected individuals from the population) and is largely worthless within families
(i.e., when trying to explain differences in siblings' schooling outcomes), then the *Bell Curve* view that genetics contributes to the
growing class divisions in society and their intergenerational reproduction seems plausible. But contrast this scenario with the case in
which a one-standard deviation change in the polygenic score between brothers results in a greater difference in actual schooling outcomes between those two siblings (rather than if we looked at two
random individuals from the population with the same one-SD difference). In that world of accentuated within-family differences, genetics may be a wash with respect to family dynasties and instead
would contribute to individual success. This might be possible in
spite of the 50 percent average correlation in sibling genetic scores
(assuming random mating). That is, thanks to genes, niche formation,
or specialization, within families may counteract the shared genetics
and lead to net social mobility.

As it turns out, we found that a given numeric difference in the
educational polygenic scores between two individuals did indeed predict a more dramatic difference in schooling outcomes if those two
individuals were siblings than if they were randomly drawn from
the population at large.[34] In other words, though siblings share their
genes to a certain extent, the differences between them matter a lot.
For each standard deviation change in the score within families,
about a third of a year of schooling difference is predicted. But when
comparing individuals from different families, the effect of the same

genetic difference is approximately one-quarter year. How can this be possible? The answer probably lies in the tyranny of small differences— that is, niche formation. Siblings are compared with each other more than with randomly selected individuals from the population. Observed differences in ability may lead parents to reinforce those differences, or siblings themselves may pursue a strategy of differentiation that leads to an accentuation of their genetic differences. The slightly more bookish one may find her identity through school achievement while the sporty sib neglects homework in favor of varsity letters. Whatever the mechanism, the effect is muting the class solidification aspects of genotypic effects (though not completely so, since the larger effect within families is not big enough to totally eliminate the impact of the shared genes between siblings). The story is more complicated than Herrnstein and Murray have suggested. Just how the balance between sibling niche formation and shared genetic ancestry plays out in terms of the reproduction of inequality across generations depends on yet another factor: to what extent parents sort on the educational genotype. Genetic sorting is the topic to which we now turn.

CHAPTER 4

GENETIC SORTING AND CAVORTING IN AMERICAN SOCIETY

In 1958, the British sociologist (and later, member of Parliament) Michael Young penned what was meant to be a satirical essay, "The Rise of the Meritocracy: 1870–2033."[1] In it, Young described a system of social sorting within the education system that was engendered by a combination of compulsory schooling laws and the expansion of civil service. The system included a fetishizing of academic degrees by the labor market, a narrow conception of skills and talent as being based on test scores, and the rise of an educational elite that works to reproduce itself. Sound much like science fiction to you? Evidently it did not to his contemporaries either, since his titular neologism, "meritocracy," was wholeheartedly adopted by scholars, pundits, and, well, basically everyone—but without the ironic, negative connotations with which Young had originally freighted the word. Interestingly, Young's tale ends with a revolt that brings down this social order not too far in the future from, oh, now.

Perhaps taking Young a little too seriously,[2] Herrnstein and Murray argued in 1994 that U.S. social policy could no longer promote economic mobility because by that point in American history (the 1990s), unfair distinctions due to social environments had been all

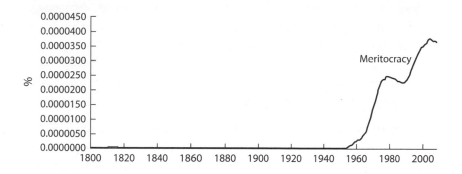

Figure 4.1. The rise of meritocracy (in print, at least). Who knows why there was a dip starting around the elections of Thatcher and Reagan? On the other hand, the second dip—after the ascendency of George W. Bush—is perhaps obvious.

but eliminated. Not that there wouldn't be differences in social environments between the slums and Park Avenue; rather, these distinctions would be the *result*, not the cause, of the economic disparities they represented. Herrnstein and Murray argued that a genetically based caste system like the one predicted by Young had already materialized in the United States by the 1990s. This genotocracy was not only being reinforced by skill-based sorting in the education system and the labor market, but also was being solidified within the process of reproduction by an increase in assortative mating on skills and intelligence, which caused the distribution of talent to widen further with each generation.[3]

In this chapter, we decide to take *The Bell Curve* seriously and test three of its core propositions empirically. We use molecular genetic data that were not available at the time Herrnstein and Murray were writing, which allow for truer tests of their own arguments. Their hypotheses are as follows.

Proposition One: The effect of genetic endowment is increasing over time with the rise of a meritocratic society:

High cognitive ability as of the 1990s means, more than even before, that the chances of success in life are good [for that individual] and getting better all the time, and these [life chances] are *decreasingly* affected by the social environment, which by extension indicates that they must be *increasingly* affected by genes.[4]

Proposition Two: We are increasingly stratifying by cognitive genotypes through the process of assortative mating:

Likes attract when it comes to marriage, and intelligence is one of the most important of those likes. When this propensity to mate by IQ is combined with increasingly efficient educational and occupational stratification, assortative mating by IQ has more powerful effects on the next generation than it had on the previous one. This process, too, seems to be getting stronger, part of the brew creating an American class system.[5]

Proposition Three: Society is adversely selecting for intelligence since those with lower ability tend to have more children than those with high cognitive ability:

The professional consensus is that the United States has experienced dysgenic pressures throughout either most of the century (the optimists) or all of the century (the pessimists).... There is some evidence that blacks and Latinos are experiencing even more severe dysgenic pressures than whites, which could lead to further divergence between whites and other groups in future generations.[6]

Herrnstein and Murray also make arguments about the role of genetic differences in explaining ongoing racial inequalities in U.S. society. We do not deal directly with those claims in this chapter but will turn to them in chapter 5.

GENOTOCRACY RISING?

Let us begin with the first proposition: that the heritability of social traits is rising over time. To assess this prospect we could take twins born in the first half of the century and those born in the second half of the century and see if we get different heritability estimates for the two groups. Or we could do the same thing using molecular data and generate heritability estimates based on unrelated individuals born before World War II and those born after. (See chapter 2 for a discussion of this method.) We have performed this very exercise for smoking behavior, for example, and found that, indeed, the genetic component explains an increasingly greater share of observed tobacco use over the course of the twentieth century.[7] In fact, it is after the surgeon general's famous 1964 report that we see the heritability rise.[8] The story here seems pretty clear: once the environmental inputs shifted with the addition of more information about the dangers of smoking, and norms started to turn against tobacco use, those with a weak genetic propensity to become addicted to tobacco quit successfully, leaving those more genetically wired to love nicotine in the dwindling ranks of smokers. Most adolescents or young adults try cigarettes at some point, and whether they develop a habit of smoking depends on both the social climate, as well as how they are wired to receive the effects of nicotine. Once the social reasons for smoking waned in importance (i.e., environmental impact), the heritability rose, with only the hard-core nicotine addicts left in the smoker pool.

It is not just novel information about the risks and rewards of a given behavior that can shift the environmental landscape. For example, should we expect the genetic influence on height to have increased over the course of recent decades or decreased? Changes in the mean (that we have grown taller and heavier as a nation, though progress on height has slowed down) and changes in the variance (that the spread of heights in the population has also increased) do not *necessarily* indicate that the heritability will increase or decrease.[9] As it turns out, in the United States the effect of genotype on height has, in fact, steadily marched upward over the course of the century, as we can see in figure 4.2. The key is the gap between the two lines in each panel. The lower line represents the predicted height by birth year for someone with a low genetic height score (−1 standard deviation from the mean). The top line is the expected stature of someone born in that year with a high predicted height based on their genetics (+1 standard deviation). If the gap gets wider from left (earlier birth

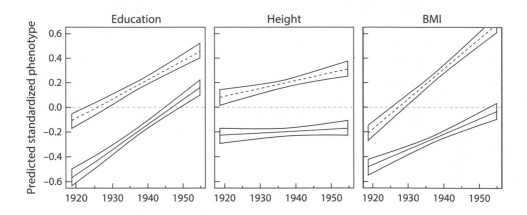

Figure 4.2. Predicted standardized values of selected phenotypes by polygenic score (+1 or −1 standard deviations), across birth cohorts among genotyped respondents in the Health and Retirement Study (*N* = 8,865). Height (*p* < .05) and BMI (*p* < .001) polygenic scores become more predictive in later birth cohorts while the education (*p* < .05) PGS becomes less predictive.

cohorts) to right (later cohorts), as it does for both height and body mass index (BMI), then the effect of genotype is increasing in more recent generations.

Of course, after looking at the chart, it makes complete sense. In a world in which nutrients are scarce, we should expect the effects of genes to be repressed. But once the environmental landscape changes so that there are no restraints on genotypic expression, our heretofore muted differences can emerge.[10] The same dynamic holds even more so for BMI. As calories become abundant and we suffer as a nation from an obesity epidemic, it turns out that a thousand genotypes bloom (and balloon) as we all reach a different genetically influenced weight-for-height.[11]

Could a similar dynamic be happening with socioeconomic status? Could the environmental landscape be shifting such that—à la Herrnstein and Murray—the restraining forces of the old world on achievement have been sloughed off and we can now all fulfill our genetic potential in the game of achievement? In the old days, when good ol' boys got gentlemen's Cs and college (or graduate) degrees were more like an accoutrement than the key to success, genotype should not have mattered much in predicting schooling compared with *social* class background. But when we entered the age of a competitive, knowledge-driven economy in which education is the ticket that must be punched, the question is whether we should expect raw talent (i.e., genotype) to be ascendant. (This is the core of the *Bell Curve* argument.)

As it turns out, the answer is no.

Genes do not predict years of schooling among those born in the 1960s better than they did for those born in the 1920s. In fact, they predict education marginally worse.[12] (As shown in figure 4.2, the gap between the two lines grows slightly narrower as we move from left to right.) This stands in contrast to some twin-based studies that had hypothesized that heritability was indeed rising.[13] If we believe that any effect of genotype is "fair" in terms of meritocracy and equal opportunity, and any environmental effects reflect unfair "noise" that we should seek to minimize, this failure of the effects of genes to rise is actually bad news. Perhaps social advantages have become cumulative over generations, and, as schooling has become increasingly important to economic security, degrees have become a cultural mechanism through which social advantage is passed on in spite of genotype. Think legacy admissions.[14]

An alternative explanation is that in a prior epoch when a high school degree was a high aspiration and any degree beyond high school was indeed a rare achievement, it really took an innate love of education to soldier on to college and beyond, especially since it was not as "necessary" for financial success and security. In this scenario, genetic aptitude for school should matter more during the earlier epoch, and contrary to the *Bell Curve* account, as we expand access to education—as we did over the course of the twentieth century—the effect of genes should wane. This is perhaps best understood by thinking through the impact of compulsory schooling laws that intend to assure that the entire population attains a minimal level of education.

If we passed a law that everyone has to attend school through age sixteen or more (as many states did), then there should be no effect of genotype whatsoever (or of environment for that matter), at least for completion of the tenth grade.[15] Supporting the interpretation that the expansion of schooling is probably driving down genetic effects on educational attainment is the case of Sweden. In social democratic Sweden, with its commitment to equal opportunity, the effect of genotype has declined over the same time period.[16] Further supporting this interpretation is the fact that when we broke out the effects on education by stage of schooling, we found that while the effect declined at lower parts of the schooling distribution (such as whether or not someone graduates from high school), we also found that the transition from a bachelor degree to a graduate degree appeared to show increased genetic predictability.[17] (Among the cohorts we studied, graduate education was extremely rare.) The bottom line is that a shifting environmental landscape can cause genetics to become either more or less salient depending on the particulars of that change. In this case, the result is that Herrnstein and Murray's proposition 1—the increasing salience of genetics to overall educational (and by extension labor market) success—does not hold up.

I, AND MINE GENES, THEE WED

But what about mating life—proposition 2? Surely we have begun to sort ourselves more by genotype in the marriage market? After all, on phenotypic measures related to social class—education, occupation, and earnings—spouses are more alike now than ever.[18] It would seem intuitive that while wooers and wooees are pairing up phenotype—since that is what they observe in each other—they are de facto increasingly matched by genotype. This may especially be the case if the phenotype is really acting as a proxy for something deeper—for example, height proxies for healthy genes or educational degree stands in for cognitive ability. So when demographers observe that a college graduate is more likely to marry another college graduate today than he or she was in 1960, this could mean that spouses increasingly value intellectual compatibility over other forms of compatibility (such as physical or religious or ethnic). Indeed, in 1960, 32 percent of men with college degrees married women with bachelor's degrees; in 2000, 65 percent of them did.[19]

Given this trend, we might expect to observe a similar increase in spousal propensities to match on the underlying genes.[20] On the other hand, we have already been surprised that the genotypic effect on education overall has not increased over time. That does not necessarily imply that spousal sorting on genotype has declined, but it may give us reason to reconsider the above narrative. To begin with, the statistic that men in 2005 with college degrees were nearly twice as likely to marry a woman who also held a bachelor's degree as men in 1960 conflates two separate dynamics. Namely, there is the simple fact that the rapid expansion of women's access to and participation in higher education over that time period has resulted in more college-educated women. Because there are more equal educational distributions between the sexes, it is easier to marry someone with the same education compared with the situation in which a significant number of men have college educations and almost no women do. That is, higher educational assortative mating becomes a mere statistical artifact that all men—not just college-educated men—are more likely to marry women with a college degree ceteris paribus. In other words, when there is a shift in the overall distributions of men's and women's educations so as to become more similar between the sexes, even random mating (i.e., no change in preferences) could generate the above-cited result. We still care about this change in the overall relative education levels of the two sexes, since it might still have consequences for marital quality and stability (for instance, if one sex rapidly increased its education while the other remained stagnant). But for our present purposes—wanting to know whether partners are sorting more now than they were before—we want to factor out the change in the relative distributions of education across the sexes and instead look at who chooses whom within each sex's educational hierarchy.

The question we really care about is whether, holding the distributions of education for both sexes constant, there is more similarity between spouses in their *relative* levels of schooling—their rank-ordered education levels. Indeed, a robust research literature has examined spousal correlations on a whole host of dimensions. Similarity (i.e., correlation) in rank order can occur as a result of at least two different dynamics. It may simply be that individuals seek out partners like themselves for a whole host of reasons ranging from shared interests and values to personality type to xenophobia. Or, alternatively, if all suitors were seeking to maximize on a given dimension—say physical attractiveness or income—then we would pair off in rank order along that dimension just like medical school graduates and residency programs match.

If the first dynamic were dominant, we would find high correlations for traits that have a clear hierarchy (say, wealth or IQ) and lower correlations for traits that do not evince obvious "better-worse" directionality (say religious preference, ethnicity, or personality type). As it turns out, the story is ambiguous. Spousal correlations tend to be relatively low for physical traits like height (0.23) and weight (0.15) as well as for personality traits (0.11 to 0.22).[21] Correlations are quite a bit higher for cognitive ability (0.40) and education (0.60)— which could be seen as evidence that we maximize when it comes to stratified traits.[22] But also displaying high spousal correlations are nonhierarchical characteristics like political ideology (0.65) and church attendance (0.71), which would seem to fly in the face of the hypothesis.[23] Of course, some of these factors—such as ideology or church attendance—may also grow more similar over the course of a marriage due to the cross-socialization effects of spouses.[24] Or it could be that we maximize on some dimensions (the smartest possible spouse regardless of our own intelligence) and match on others (e.g., religious preferences) and leave still others to chance (height, perhaps).

To try to understand this by ruling out cross-socialization over time, a pair of scientists randomly manipulated political information contained in online dating profiles. The authors found that subjects participating in the experiment rated profiles more favorably if those profiles described an individual who shared their political ideology.[25] The study also found, based on an actual national online dating community that men were more likely to send a message to women with shared political traits. Reciprocally, women were more likely to respond to a man's message if they had political traits in common.[26] A practical implication of this research is that liberals tend to marry and have children with other liberals and conservatives tend to marry and have children with other conservatives. This positive assortment leads to more polarization in political attitudes among the overall population and pushes individuals out to the liberal and conservative extremes. In other words, positive assortment may contribute to political polarization.

While the marriage market may be part of the explanation for contemporary American political cross fire, it does not appear to matter as much for changes in class structure. The fact that highly educated, high-earning men increasingly marry highly educated women does, in the cross section, explain some of our income inequality in the United States (although not as much as it would if women were participating in the labor market to the same extent as men, thereby maximally cashing in on their educational returns). Over time, it turns out that the continually rising returns on education—for both sexes—are much more important than educational sorting in explaining the rise in household income inequality in the United States (or Norway, for that matter). In fact, while there has been an increase in sorting at the bottom end of the educational ladder (i.e., those with less than a four-year college degree more and more tend to consider education level in pairing up), there has actually been a decrease in the matching at the top end (i.e., among college graduates), measured by either level of schooling or major between 1980 and 2007.[27] The end result is that the spousal correlation has barely budged and, by some estimates, may even have declined.[28]

It is not a crazy idea to think that in the face of a more unequal economy, spousal similarity may have declined. The economist Gary Becker suggested a model of the family in which specialization between spouses—one focuses on bringing home the bacon and the other on frying it up—would lead to negative assortment on some traits. That is, if small differences in education level, skills, or raw ability lead to huge differences in the realized economic returns in the labor market, it could mean that a better mating strategy would be for each spouse to specialize in an area of expertise. Someone who is highly cognitively endowed and able to earn lots of money could choose a stay-at-home spouse who contributes other abilities to the household, such as empathy and care-taking ability.[29] Of course, whatever we observe with respect to the correlation of education levels or even of test scores between spouses, this does not tell us what may be going on beneath the surface—that is, at the genetic level.

How would we, in fact, assess whether spouses are more genetically alike than random strangers? It is not as straightforward as measuring phenotypic assortative mating, and we have already seen that even just assessing educational assortative mating requires decisions on what we want to estimate. We decided to tackle the issue of genetic assortative mating (GAM) in several ways. First, we looked at overall genetic similarity. We found that in general, spouses were definitely more genetically similar to each other than randomly paired individuals in the population. Our spouses were, on average, not quite as genetically similar to us as a first cousin once removed, but they were more similar to us than a second cousin. While this analysis was restricted to whites, we found that even when we factored out historical patterns of marriage within white ethnic groups, we still observed genetic relatedness among spouses equivalent to second-cousin status (between 2 and 3 percent genetic similarity). Indeed, being a standard deviation more similar genetically increased the probability that you would be married to that person by 15 percent!

Before you say, "eww," keep in mind that this analysis was conducted in the United States, which is a highly mobile, immigrant society. That is, we probably have one of the lowest genetic relatedness coefficients for spouses in the world. In a tribal society like Pakistan, for instance, the rate of first-cousin marriages is well over 50 percent. The lack of long-distance migration around the world has also likely produced relatively high genetic similarity between spouses. Until relatively recently, perhaps a century ago, evidence suggests that many families remained in the same area for generations, with men courting no more than about five miles from home—the distance they could walk out and back on their day off from work. As a result, it's likely that 80 percent of all marriages in history have been between second cousins or closer.[30] We are essentially showing that though we no longer explicitly marry our cousins in contemporary

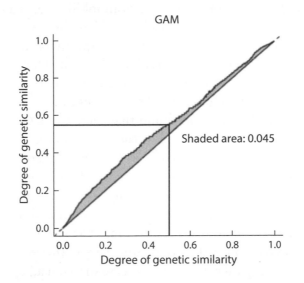

Figure 4.3. Spouses are related at a degree of .045, equivalent to first cousins once removed in the United States. When ethnicity is factored out through controls for principal components, the level of genetic similarity between spouses falls to that of second cousins. The *y*-axis charts quantiles of the distribution of kinship between all pairs of observations. The *x*-axis charts quantiles of the same distribution but restricted to just cross-sex white spousal pairs. The shaded area gives an estimate of assortative mating. The horizontal and vertical lines aid in interpretation. Domingue, BW, Fletcher, J, Conley, D, and JD Boardman. (2014) Genetic and Educational Assortative Mating among US Adults. *Proceedings of the National Academy of Sciences* 111(22): 7996–8000.

U.S. society, from a genetic point of view, the results are not all that different.

The genetics of marriage is not so surprising when we consider that friends are also more genetically similar than random strangers. Nicholas Christakis and James Fowler found that friends (who were not relatives) were genetically the equivalent of fourth cousins.[31] Interestingly, friends in their network study shared more SNPs than nonfriends, but they also displayed an excess of "opposite" genotypes, loci that diverged more than they would by chance. When Christakis and Fowler investigated these respective patterns of homophily and heterophily, they found that the homophilic (i.e., like-likes-like) genes tended to cluster in two biological pathways: linoleic acid metabolism and olfactory (i.e., smell) perception. Now, who knows why linoleic acid metabolism would be a bonding glue for friends?[32] But olfaction makes sense. We like people who have the same taste in smell as we do (so much for mixed song tapes ...).

What's more, these authors found that there was a certain class of genes that were overrepresented in the heterophilous group: those related to immune function. This has long been theorized with respect to spouses, who are hypothesized to be discordant in their genotypes for a particular region of chromosome 6 that codes for immunological genes called the major histocompatibility complex, or human leukocyte antigens (HLA) area. The theory is that evolutionary forces have pushed us to seek diversity in this genetic package that confers biological resistance to disease so that if an epidemic hits our family or wider tribe, at least some individuals in the group will have native resistance and survive. In other words, it is an insurance mechanism of sorts. If it works for spouses (supposedly through smell, actually), it should also hold for our wider group of associates.[33]

While the global similarity of spousal (and friend) genotypes is intriguing and suggestive, it does not really answer our question about sorting on specific genotypes that might lead to genetic castes in society. To answer that question we really need to know whether spouses are sorting on the genes that matter to particular outcomes. Twin-based heritability estimates, adoption studies, GCTA models, and sibling IBD approaches are all indirect methods that tell us about how much phenotypic variation is accounted for by genotypic variation in the population. But they tell us nothing about how alike two given individuals are on the genetic variants that matter. Nor do such methods allow us to see how a measured genetic effect may be interacting with a measured environmental influence (the topic of chapter 7). In order to do that, we need to look at specific markers, using the polygenic score approach outlined in chapter 3.

There is certainly reason to suspect that sorting on particular genotypes plays a large role in the mating markets of contemporary society. When we plot spousal correlations on outcomes for which we have decently predictive polygenic scores from major consortia (education, height, BMI, and depression) in figure 4.4, we find that, with one major exception (height), social sorting dwarfs genotypic sorting, at least by these measures. When we look at a phenotype that potential spouses can observe at the time of making their mate choice, such as education, we find a high correlation in phenotype; that said, this does not hold true for height, which is also observable at the time of marriage. The other two outcomes we look at in terms of spousal similarity—BMI and depression—are arguably only semi-observable at the time of marriage. Spouses may not have fully ballooned into the BMIs we observe in the Health and Retirement Study survey when most are over 50 years old, but they have probably shown some tendencies to be fatter or thinner by the time they marry. Likewise, depression rates increase with age, but the inklings of depression may be present at the time of marriage. So they show moderately strong correlations. And in all these cases (height excepted[34]), spousal similarity on the underlying genes is substantially lower than it is for the actual outcome that the genes are meant to predict. This makes sense, since we are not (not yet, at least) genotyping our potential spouses and deciding whether to marry them based on the results. We are observing their outcomes on the dimensions that matter in society. Plus, we must recall that these polygenic risk scores are noisy proxies for the entire genetic component of these traits.[35]

So while we are indeed very correlated with our spouses in level of education, most of this seems to be social sorting rather than genetic sorting—at least based on the measures we have.[36] More important than the absolute level of genotypic correlation to the *Bell Curve* argument is the trend. In figure 4.5, we show the trends in the similarity between spouses on phenotypes and genotypes. In these charts, the two lines represent the predicted education (or height or BMI or depression) level for the spouse of someone who is at the upper (+1 standard deviation) or lower (−1 standard deviation) parts of the educational (or height or BMI or depression) distribution. So a larger gap means more assortative mating; a smaller gap means that your own education or BMI or height or depression score does not predict as much about your spouse's.

When we examine the phenotypic data in panel A of figure 4.5, we indeed reproduce what demographers have been finding without

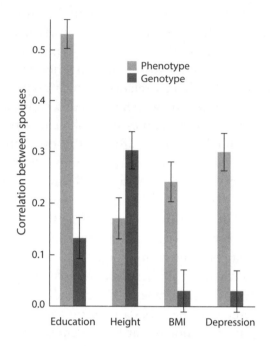

Figure 4.4. Spousal correlations in selective genotypes and associated polygenic risk scores among married couples in the Health and Retirement Study, 2012 (N = 4,909). Restricted to respondents in their first marriage who have genotypic data and valid phenotypic responses. Height is the only trait for which the phenotypic spousal intraclass correlation is significantly lower than the genotypic intraclass correlation. Polygenic-score-residualized phenotypic correlations are indistinguishable from raw phenotypic correlations, suggesting two distinct sorting processes at work.

genetic data: spousal similarities on education are higher for couples born later in the century than those born earlier in the century. That is, the gap widens between more highly and less-educated individuals (and more- or less-depressed individuals) in terms of the education (or depression) level of their spouse (while trends are flat for BMI and height). This would appear to provide support for the Herrnstein-Murray argument. However, when we look at genotypic similarity, much to our surprise, spousal correlations are flat. So much for the *Bell Curve* argument that we are increasingly sorting by genotype.[37]

A

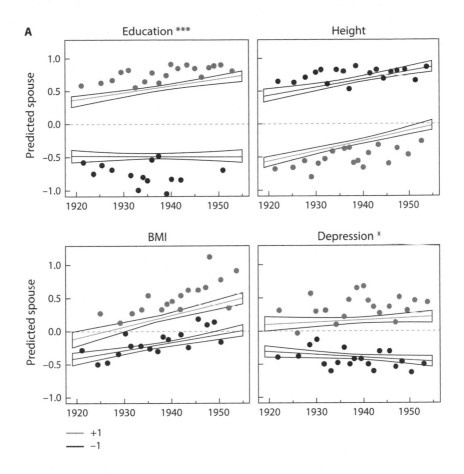

Figure 4.5. Cohort differences in spousal correlations in selective standardized phenotypes (panel A) associated, standardized polygenic risk scores (panel B) among married couples in the Health and Retirement Study, 2012 (N = 4,909). Restricted to respondents in their first marriage who have genotypic data and valid phenotypic responses. Confidence intervals for difference in correlation estimates computed by inverse Fisher z-transformation. Asterisks and plus signs indicate statistical significance of birth cohort X genotype interaction. + *p* < .10, * *p* < .05, ** *p* < .01, *** *p* < .001.

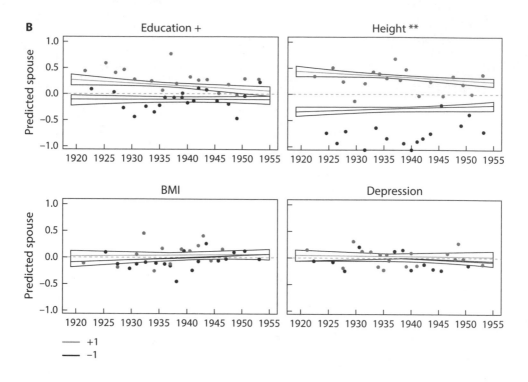

Figure 4.5. (continued)

So we feel safe in at least saying that there is no convincing evidence *for* the Herrnstein-Murray argument that we have slipped into the brave new world of genetic castes reinforced through marriage dynamics. At least not yet.[38]

IDIOCRACY, HERE WE COME?

There is one last genotocratic-related question to address and this one is more about where we are going as an entire society than who gets ahead within a given generation. Namely, are we becoming less cognitively endowed over time thanks to differential fertility by genotype? While this is not a central *Bell Curve* hypothesis, Herrnstein and Murray do raise this possibility, and others have more forcefully suggested that differential fertility by social position can have huge effects on the population distribution in subsequent generations—and even on the economic fortunes of an entire society. The economic historian Gregory Clark, for instance, argues that an overlooked determinant of the Industrial Revolution's origin in England was the higher fertility rates of wealthy (read: smart and capable) citizens compared to the lower classes. Never mind natural resources such as coal deposits near the surface or a stable set of institutions (see chapter 6 for a discussion of the various causes of economic growth), Clark claims in A Farewell to Alms that the key variable was the genetic stock of the population. In preindustrial England, those who were more economically productive also had more surviving children than those who were less economically successful. Over time this led to a "bettering" of the genetic stock, as with dog, apple, or livestock breeding. (By "bettering," he means more adapted to what would become modern economic life.) At a certain tipping point, productivity exploded and voilà—we get the rapid rise in incomes and population growth that we have witnessed over the past few centuries after millennia of stagnation.

Whether or not we believe Clark's story, a common (if mostly unspoken) concern of certain right-wing psychologists and geneticists is the flip side of this pro-development fertility and survival bias—what they call dysgenics: a reproductive pattern in a population that favors those with undesirable traits (like low education) over those with more desirable features.[39] Namely, if it was the rich having proportionately more kids than the poor that helped us all get richer, then should we not worry if the reverse is true today—that is, the less educated are having more kids than the better educated or cognitively endowed are? One consequence, after all, of what social scientists call the demographic transition is that not only does everyone have fewer kids, a higher percentage of which survive, but also that many of the predictors of fertility (such as a woman's education level) reverse their impact. For example, whereas in the high-fertility, high-child mortality regime of yesteryear (which is still ongoing in many less-developed countries today), more educated people had a greater number of (surviving) offspring, today it is the case that poorer families have more children. In our analysis of the Health and Retirement

Study, we find a correlation of –.18 between years of schooling and number of offspring. Other studies have found that women with a college degree have, on average, one fewer child in the United States than those with only a high school degree.[40] The negative relationship between female education and number of kids is so strong that a clever paper by the demographers Robert Mare and Vida Maralani showed that raising women's education levels in one generation does not necessarily translate directly to higher education levels in the next generation for the population as a whole. That is, the education-related reduction of fertility is so strong that the women who obtain more schooling end up having fewer kids. Even though the children they do bear tend to have higher levels of education themselves, that effect is partially cancelled out within the population as a whole by the fact that a greater share of the population is born to mothers who are in the low-education group, since they do not experience the same fertility reduction.[41] Are we, in fact, breeding ourselves into a low-education population? And how does this jive with the well-noted Flynn effect, according to which measured IQs have been rising during the past century?

When in panel A of figure 4.6 we look across cohorts at the correlation at the *phenotypic* level—the actual educational level of a respondent and his or her number of children—we find the pattern expected from the demographic research. Those born before 1940 show basically no correlation between education and the number of children. (These parents were born in the middle of the demographic transition in the United States.) And for those born 1940 or later, there is a significant, negative relationship between years of schooling and fertility. (The correlation is –.20.)

As you may have guessed, to determine if we are breeding ourselves into a less intelligent population, we need to parcel out the education-fertility association into its genotypic and phenotypic parts. There does appear to be a stably heritable component of fertility: the additive heritability for number of children has remained consistent across cohorts at around 20 percent. Clearly, fertility is somewhat genetically influenced—despite some theories that suggest that the long-run heritability of fitness should be zero. It is therefore plausible

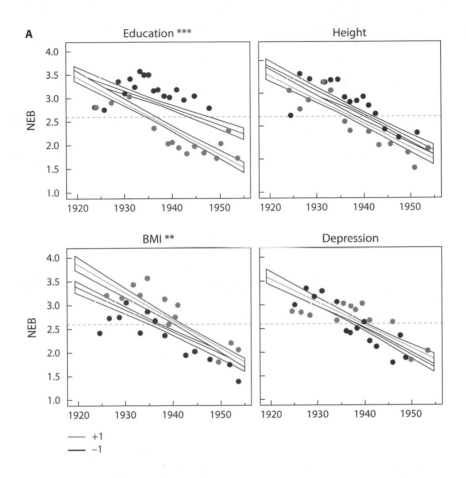

Figure 4.6. Changes in the association between selected phenotypes (panel A) and polygenic scores (panel B) and number of children ever born (NEB) across birth cohorts in the twentieth century among genotyped Health and Retirement Study respondents (N = 8,865). (A) All phenotypes show changing associations with fertility. Education shows bigger gaps in fertility among the youngest birth cohorts, while body mass index (BMI) shows a decreasing association with NEB. (B) Despite changing phenotypic correlations with fertility, genotypic associations appear to be constant across birth cohorts. Asterisks and plus signs indicate statistical significance of birth cohort X genotype interaction. $+ p < .10$, $^* p < .05$, $^{**} p < .01$, $^{***} p < .001$.

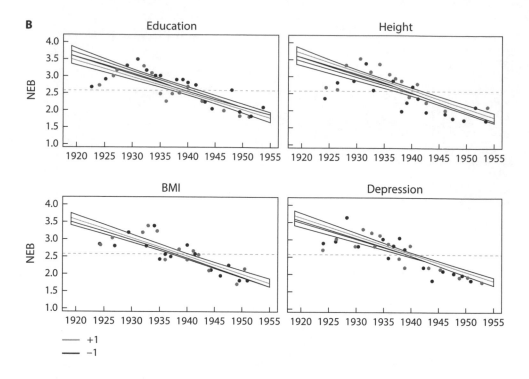

Figure 4.6. (continued)

that other genetically influenced traits (like educational ability) may share some common genetic roots with fertility.[42] But when we examine the correlation between the polygenic score for education and number of offspring, we find it to be –.04. Not zero, but close to it. Furthermore, when we look over time, in panel B of figure 4.6, we do not find evidence of increasing negative selection on education—that less educationally advantaged individuals are having increasingly more children than those who are more educationally endowed, genetically speaking.

While it is true today that the higher up the education ladder one climbs, the fewer children that person is likely to have, this relationship appears to be mainly related to the social roots of schooling and not to the genetic ones (at least as proxied by the education genetic score). More critically, the small negative correlation between education genotype and the number of children is not increasing even as the corresponding relationship between measured education and the number of children is getting stronger and stronger. In other words, there is a lack of evidence to support the notion that differential fertility by education level is leading to a situation of negative *genetic* selection on education. Fans of the 2005 movie *Idiocracy* and their dysgenicist comrades can breathe a sigh of relief.[43]

THE BIG SORT

We are left with a puzzle of sorts: How can spouses be so similar (phenotypically) and yet so dissimilar (genotypically) at the same time? And as spousal phenotypic similarity increases over time, why do we find no subsequent genetic similarity increase? The logic motivating these questions is easy to follow: (1) spouses are very similar on many measures, like educational attainment (2) these measures show moderate heritabilities and, thus, (1) + (2) → (3) spouses should be genetically similar.

While spouses are indeed *somewhat* similar genetically, there is no trend of *increasing* similarity. One may conclude, then, that the observed increase in educational sorting noted by many scholars is happening on the environmental side. This is a very interesting social "fact" in its own right. One obvious explanation is that men, but especially women, have increased their schooling to a vast degree over the course of the twentieth century. (This gender difference is perhaps part of the reason why when the original GWAS analysis that created the polygenic score was run separately by gender, the genes that seemed most predictive appeared to be different for the two groups.[44])

We may have always been sorting on academic "ability" to a certain extent—at a level that has been flat or slowly declining—but now with women able to earn degrees, sorting on observed academic degree is increasingly aligning with the mental qualities on which spouses "clicked" before. In earlier days, the college-educated man may have married a high-school-educated woman, but he selected the most quick-witted, erudite one (perhaps even shunning the dimmer woman who did attend college thanks to family resources and the like). Today, intelligent men and women are both much more likely to actually obtain college (or higher) degrees, leading to a rise in educational sorting even as pairing up on the underlying, relevant genotype does not rise.

The substantial differences in genetics between spouses suggests a large role for genetic mobility across generations. While meritocratic environments may be a powerful force toward aligning successful parents, their children are faced with a genetic shuffling that helps to level the playing field in the next generation—somewhat mitigating a march toward a solidified genotocracy across generations. These children are still quite advantaged, of course, in the cross-section due to both environmental and genetic factors.

The Bell Curve argued that we have achieved the societal situation in which the result of meritocracy and assortative mating is a system of class stratification based on innate (i.e., genetic) endowment, as first satirically proposed by the sociologist Michael Young in 1958 when he coined the term "meritocracy." If this were true, social policy to promote equal opportunity would be counterproductive—at least the grounds of efficiency—because each individual would have already reached the level of social status best suited to his or her innate abilities. Meanwhile, by selectively breeding with others of similar genetic stock, parents would reinforce their offspring's advantages or disadvantages. Such a situation would call into question the notion that intergenerational correlations in socioeconomic status variables—such as income, occupation, or education—are indicative of a nonmeritocratic society.

Herrnstein and Murray made their assertions in 1994, before the human molecular genetics revolution took place. Whatever one believed about their assertions or the politics thereof, they were largely untestable at the time. They based their claims on an analysis of cognitive ability, which is problematic because IQ has both environmental and genetic bases, so any trend in its effects could be attributable to the environmentally influenced portion or the genetically determined one. Furthermore, they analyzed the National Longitudinal

Survey of Youth 1979 (NLSY79)—a survey of men and women born in the years 1957–64. This is hardly a wide enough swath of birth cohorts (particularly since everyone was born post–WWII) to test their grand theory. In contrast, we tested their hypotheses using molecular data across a much wider (and frankly, more appropriate) birth-cohort distribution. Though we are a mere decade and a half away from the 2033 date on which Young predicted the final "revolt" against an entrenched meritocratic system (in the U.K.) would occur, it seems—if the present results are to be believed—that we are still considerably removed from the dystopian nightmare he imagined more than half a century ago.

But what might happen if the process of (parental) phenotypic sorting and (child) genetic shuffling morphed? What if would-be spouses were not limited to relying on the vague and downstream signals of educational attainments and earnings in deciding with whom to partner and instead could match on genotype? What if eHarmony and OkCupid merged with 23andme, and a key feature of your profile was your educational polygenic score rather than your actual academic degrees? We explore this possibility in the conclusion to this book.

CHAPTER 5

IS RACE GENETIC?

A NEW TAKE ON THE MOST FRAUGHT, DISTRACTING, AND NONSENSICAL QUESTION IN THE WORLD

When I (Conley) had my entire family genotyped, many of the results could have been predicted just by looking at our phenotypes. For example, my sporty ex-wife had two working copies of the so-called sprinter's gene, while I had none. But there were also some surprises in store for us in the area we tend to take the most for granted as "obvious"—race. I had long known that I was half Ashkenazi, but my family had debated whether the other 50 percent was part Native American. My father's X chromosome turned out to have a haplotype that was present in pre-Columbian North America. So family legend that we are part Mashpee Indian survived. The biggest shock was that my ex-wife was a quarter Ashkenazi herself. Further digging revealed that her father's father, was, in fact, 100 percent Jewish, ethnically speaking. A Ukrainian man who could trace his lineage back to the Cossacks and whose attitudes toward Jews betrayed the history of mistrust in Eastern Europe, he would have to rethink his entire world view about race and politics.

In ways small and large, genetics has become an important new tool in revealing the deep past for scholars and family-tree buffs alike.

Genetics holds the promise of complicating racial narratives not just within individual family units, but for entire societies. Take the case of Ecuador. That the South American country enjoys a mix of populations has been well known. There are the indigenous peoples whose ancestors crossed the Bering Strait thousands of years earlier. There are the Ecuadorians of African descent whose ancestors were brought west during the slave trade, and there are those descended from Spanish imperialists. But what exactly did mating look like over the course of post-Columbian history?[1] Analysis of DNA tells us not only the relative proportions of those three groups within the population (indeed, within specific individuals), it can even tell us about the nature of sexual interaction. For instance, while the proportion of European ancestry within the population was 19 percent, the proportion of Y chromosomes that were of Spanish origin was 70 percent. This asymmetry is, of course, not surprising. Most of the Spanish who came to run the Encomienda system in the New World were men. And genetics reveals the brutal nature of male European rape of indigenous women—a fact well known through traditional historical methods.

However, what about African–Native American relations? Little is known in the historical record about these interactions. Geneticists found that among self-identified Afro-Ecuadorians, the percentage of Native American ancestry was 28 percent, while the corresponding figure for the Y chromosome was 15 percent (the highest for any African-derived American population so far analyzed). This finding suggests that there was significant reproductive mixing between the two populations and it mostly involved black men with Native women (though the reverse was not uncommon). This result does not close the book on Ecuadorian reproductive history, but it certainly adds a data point that should be incorporated into the more traditional scholarship of Ecuadorian race relations—since sexual contact is an important indicator of racial power dynamics, degree of assimilation, and so on. For us, this study prompts the broader question of what genetic analysis can and *cannot* offer in terms of understanding race and racial history in the United States.

As we cross into this terrain, it is worth keeping in mind that the intersection between race and genetics is probably the area of scientific and pseudoscientific analysis that has done the most damage in our collective history. One difficulty in discussing these topics is that there are not only actual important scientific questions at play, but also awful ideological traditions of using scientific jargon to subjugate people in the name of science, fate, and the natural order.

INTERROGATING WORLD HISTORICAL GENETIC PROCESSES WITHIN THE PRESENT-DAY UNITED STATES

In the United States, we tend to classify someone as black based on the rules of hypodescent—that is, the one-drop rule, according to which any amount of African heritage makes someone black. We further divide the nonblack population into Asian, Native American, and white races. In terms of Native American ancestry, we generally use a rule of hyperdescent—that is, those Euro-Americans who have a small amount of Native American ancestry are generally considered white. In contrast to the approach to race, many population geneticists use cladograms—family trees, of a sort—to represent the genetic distances between peoples. To construct a cladogram, the researcher measures the evolutionary time since divergence from a common ancestor for two groups by tracing the distance up the branches to the common node. This evolutionary time is also a proxy for genetic differences—the longer the time since the split, the more time there has been for distinct sets of novel mutations to occur in the two groups (although genetic differences are affected by genetic drift as well[2]).

If we examine the unrooted genetic tree mapped by Sarah Tishkoff and colleagues shown in figure 5.1, we see that the U.S. "official" race categories do not align with the genetic differences found in our populations. If race categories were meant primarily to capture differences in genetics, they are doing an abysmal job. For example, we can see that the genetic distance between groups *within* Africa is as great

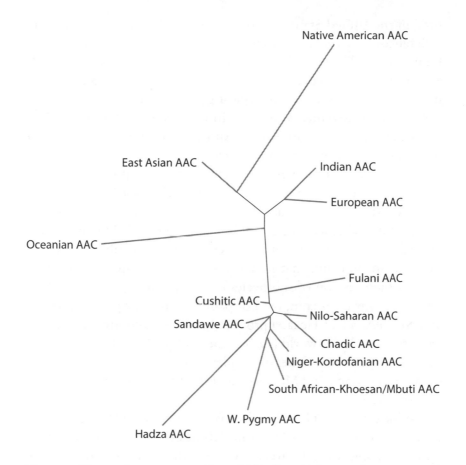

Native American AAC

East Asian AAC

Indian AAC

European AAC

Oceanian AAC

Fulani AAC

Cushitic AAC

Sandawe AAC

Nilo-Saharan AAC

Chadic AAC

Niger-Kordofanian AAC

South African-Khoesan/Mbuti AAC

W. Pygmy AAC

Hadza AAC

Figure 5.1. Unrooted ancestry tree. (From Tishkoff et al. The genetic structure and history of Africans and African Americans. *Science* 324, no. 5930 [2009]: 1035–1044.)

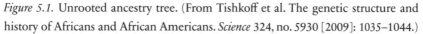

as the genetic distance between some very "racially divergent" groups in the rest of the world. If we trace the path from East Asian to European, we can see that our finger traverses a distance that is shorter than if we march along the branches from the Hazda in north-central Tanzania to the Fulani shepherds of West Africa (who live in present-day Mali, Niger, Burkina Faso, and Guinea). This means that East Asians and Europeans are more genetically similar than the Hazda and Fulani are. The fact that the racial and ethnic categories typically used in the United States do not map *particularly well* onto genetic differences between groups should be of no surprise to those who know the racial history of the United States. If we started from scratch and wanted to genetically label groups in the United States, how might we proceed? Based on typical genetic differences, we should lump "whites" together with Asian Indians. And perhaps group Native Americans with East Asians. And we certainly would want to have more than one category for "blacks."

In the modern time period, what we call race probably started with the work of Johann Friedrich Blumenbach in the late eighteenth century.[3] Much scientific and legal discussion at the time was centered around whether "negroes and savages" were fully human or part-animal—a question asked at the species level. Blumenbach's innovation was to proclaim that humans were a single species and then address the question of subspecies classification (races). In 1775 he proposed four races composed of people from Europe, Asia, Africa, and North America. He later proposed a fifth race comprising people from the "southern" world (such as the Philippines). By 1795 he had named five generic varieties of people: Caucasians, Mongolians, Ethiopians, Americans, and Malays—labels that have somewhat survived into the present day.[4]

Fast forward a couple of hundred years, and most U.S. social scientists are still using similarly murky definitions, usually defining race by how people answer survey questions that ask them to indicate whether they are white, black, Asian, Native American, or other. (Hispanic origin is asked in a separate question.) The number of potential categories someone can select has grown (mixed race, Pacific Islander, and so on), but the exercise is the same: most people select one of the socially constructed choices of race in front of them that maps onto their skin color, ancestry, or self-conception. Meanwhile, many investigators in the biological sciences have replaced the term "race" with the term "continental ancestry." This in part reflects a rejection of "race" as a biological classification. For example, every so-called race has the same protein-coding genes, and there is no clear genetic dividing line that subdivides the human species. Another reason for using the term "continental ancestry" in lieu of "race" is improved precision for locating historical and geographic origins when we look at the genome. Thus, continental ancestry allows for more genetically accurate descriptors; for example, President Obama was not just the first socially "black" president. He was also the first (as far as we know) who has European and African ancestry. With detailed data, someone can ascertain her relative proportions of Finnish and Irish and Quechua ancestry.[5] While we typically use race and ethnicity as shorthand for ancestry in our parlance, there is no reason why we should continue to do so once we have access to genetic data. Indeed, personal use of genetic data will likely overturn the incorrect folktales passed down in our families that both simplify and bludgeon the complex narrative of human amalgamation recorded in our DNA.[6]

It becomes all the more ironic that the racial-ethnic classification system in the United States has sought to destroy "ethnic" identity for the racial group in which such differences actually have the greatest genetic significance—African Americans. Although other groups— Latinos, Native Americans, Asians, and whites—have ethnic groups within their ranks (in the case of Native Americans this is explicitly a nation or tribe, whereas with the others it is usually associated with a country of origin), black Americans have no ethnic subgroups (though with increasing numbers of immigrants from majority black countries, black ethnicity has reemerged as hyphenated Jamaican-Americans, Nigerian-Americans, and so on). This happened because slave owners deliberately mixed slave populations from different geographic origins in order to break down solidarity among the enslaved and thus prevent revolt. The mélange of tribal-ethnic origins of African Americans, combined with the complete separation from their land of origin and the need to construct new cultural practices once here (and adapt old ones), meant that African Americans were effectively stripped of their "ethnic honor"[7]—that is, the pride of belonging to a group with its own history, traditions, and nationhood that exists outside the borders of the immigrant society that forms the U.S.[8] Blacks do not have the equivalent of St. Patrick's Day or Cinco de Mayo.[9] Indeed, if U.S. celebratory holidays were allocated based on genetic distinctiveness, we would have multiple holidays for each of the several tribes in Kenya and drop St. Patrick's Day altogether. How might a more accurate racial/ethnic categorization system work if it focused on genetic difference as a key to population grouping?

MEASURING GENETIC DIVERSITY

The brief history of humanity goes roughly like this. According to the fossil record, about 200,000 to 300,000 years ago, somewhere in the Rift Valley in the northeastern part of sub-Saharan Africa, the first *Homo sapiens* appeared. Our nearest living relative today is the chimpanzee, with whom we share a common ancestor approximately 5 million years ago. There were a lot of intermediate species that died off, including *Australopithecus*, *Homo erectus*, and, of course, *Homo neanderthalensis*, who may not be as extinct as we think, since some of us have Neanderthal ancestry.[10] About 100,000 years ago, Neanderthals ventured out of East Africa and headed north (figure 5.2). This was one of many migrations out of Africa. Other species that either left Africa or evolved from the groups that left Africa include *Devonians* and *Homo heidelbergensis*.

For about 40,000 years, these earlier, non–*Homo sapiens* species had the Eurasian continent to themselves (at least with respect to hominids). But about 100,000 years ago, a group of *Homo sapiens* left Africa and fanned out through the rest of the world (though some scholars think there were multiple such migrations). Evidence suggests that the effective population size (i.e., the mating pool from which others have descended) for this group was as low as 1,000 to 2,000 individuals. It is not clear whether this modest society was the result of a small number of initial pioneers who inched their way north from the Horn of Africa or whether the travails of migration across harsh environments killed off many of a larger number who had attempted to leave. Either way, the key fact is that all humans not of direct African descent share these 1,000–2,000 ancestors.[11]

This "population bottleneck" during the migration out of Africa resulted in the most fundamental genetic cleavage among present-day populations, which is with respect to continental origin: differ-

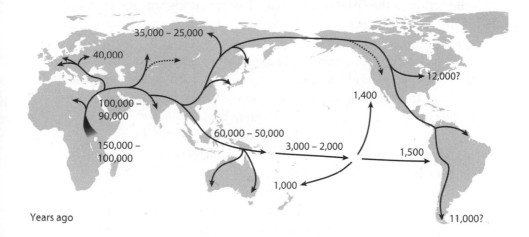

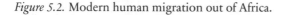

Figure 5.2. Modern human migration out of Africa.

ences in the amount of genetic variation between Africans and non-Africans. When there is a small effective mating population, some genetic variants are likely to die out through "fixation." That is, absent selective pressures, the neutral theory of evolution suggests that genetic markers rise and fall by chance in a population.[12] If a population is large, it is unlikely that a given marker will disappear by chance in a given generation. But in a small population, it is much more likely to happen. Think of a SNP that is present in 10 percent of the population. If your reproductive pool contains 10 people, each with two chromosomes, there are just two copies of the marker. It is very possible that in a given generation, neither of those two may be transmitted—specifically there is a one-in-four chance. However, in a population of 100, the chances that none of the 20 copies of that particular polymorphism get passed on are very slim: 1 in 10 million. Thus, small populations of out-migrants from Africa leave the continent with only a subset of variation and are apt to "lose" some of this remaining variation over time. Both forces result in a pruning of genetic variation the farther we travel (by foot) away from Africa, as the founding populations become smaller still. Indeed, as humans fanned out across the world, genetic diversity declined among those who ventured the farthest. Think of it like spreading jam across a piece of bread—as you scrape the knife's edge across the surface, you are spreading less and less, resulting in a very thin coat by the time you reach the other crust. Similarly, migratory distance from East Africa becomes a very good proxy of the genetic diversity in a population.

There are many ways to measure genetic diversity, but the simplest is the rate of heterozygosity (the frequency with which individuals do not have the same alleles at the same locus on both chromosomes). If a particular polymorphism is evenly split between the two alternate alleles in a population, the rate of heterozygosity will be high (50 percent—assuming random mating[13]), but if the prevalence of one of the two alleles approaches zero, the proportion of individuals who will be heterozygous at that locus is practically none. Another way to assess genetic diversity in a group is to examine the variation in the number of times a given sequence—say AGGTCT—repeats in a row, which is called a copy number variant (in a short tandem repeat).

We can see in figure 5.3, which plots SNP heterozygosity against copy number variation, that the two measures track each other very well. Furthermore, we can see that—with some exceptions—the African groups cluster at the top right of the graph (lots of diversity), while the Native American and Pacific Islanders are in the lower left (limited diversity), which is what we would expect based on their respective distances out of Africa.[14] The lower panel shows an enlarged version of the top panel, focusing on the upper right portion of the graph. Note that some of the "tribes" labeled in this illustration include "North Carolina" and "Pittsburgh," which display high levels of genetic variation. These are samples of African Americans from these localities, demonstrating that despite admixing with Europeans and Native Americans, and despite the potential founder effect (i.e., bottleneck) of the Middle Passage in the slave trade, black Americans have preserved a high level of genetic variation. This greater genetic variation among those with African origins (compared with those who left Africa by foot) reinforces that the main cleavage between high and low genetic diversity is determined by African origin.[15] In sum, racial categories now in use are based on a

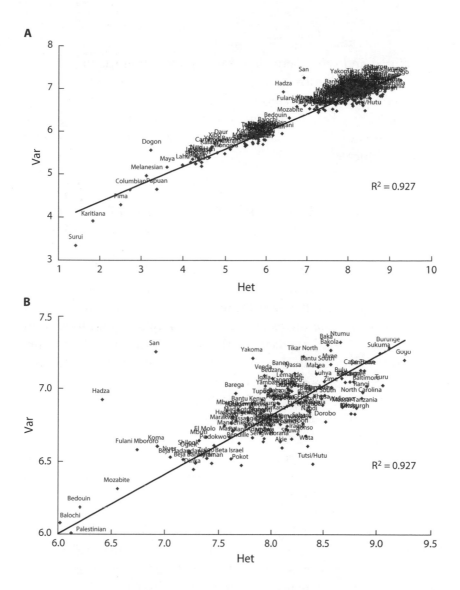

Figure 5.3. Measures of genetic diversity plotted against each other: SNP heterozygosity (*x*-axis) against copy number variant variance (*y*-axis) for world populations (A) and for a subset from upper right quadrant of world population plot (B). Those in the upper right in panel A have the greatest level of genetic diversity (groups who tend to be close to the cradle of human origins in East Africa). In the lower panel, we see that those with African ancestry in the United States are among the groups in the upper right quadrant from panel A. (From Tishkoff et al. The genetic structure and history of Africans and African Americans. *Science* 324, no. 5930 [2009]: 1035–1044.)

convoluted and often pernicious history, including much purposefully created misinformation. Even if five groups was a reasonable
first approximation, the categories we use would not satisfy any scientific criteria based on maximizing genetic difference between groups.
This is true even if we could measure these differences accurately,
which is difficult.[16]

DO ANCESTRAL DIFFERENCES MATTER,
OR ARE THEY JUST BACKGROUND NOISE?

Race does not stand up scientifically, period. However, continental
ancestry is, in fact, genetically informed. But what do these continen-
tally based genetic differences mean in modern society? It is a good
time to dispel some myths about genetic variation that have been
promulgated by both the Left and the Right alike. On the Left, many
persons try to discredit the notion that genetic variation underlies
group differences by pointing out that there is more genetic variation within these groups than between them. Another favorite approach is to cite the fact that *all* humans are 99.9 percent genetically
identical and that no group of humans has a gene (i.e., a coded-for
protein) that another group lacks. Both of these arguments are ca-
nards. After all, we are also 98-plus percent identical to chimps and
99.7 percent similar to Neanderthals. Oh, what a difference that 2 percent (or 0.3 percent) makes!

Simply stated: Overall genetic variation tells us less than specific
differences that matter. Imagine a group of humans that had a mutation in the *FOXP2* gene—often called the language gene—such that
this transcription factor (a gene that helps stimulate the expression
of select other genes) was nonfunctional. These humans would lack
the ability to communicate through language. In fact, this gene's significance was first discovered through the study of an English family
in which half the members across three generations suffered from
severe developmental verbal dyspraxia—they could not communicate orally. This family could be 99.9999 percent genetically identical
to their neighbors, but what a huge difference that 0.00001 percent

makes. This criticality of particular genetic differences, as opposed to global similarity, is not unique to humans. Through genetic manipulation of just four genes, scientists in the lab have been able to turn a mustard weed into a woody tree. It sounds like a genetic version of the 1970s game show, *Name that Tune*: in how few notes (or genes) can one radically alter the phenotype of an organism? (Keep in mind that while common variants like SNPs have very small effects in the general population, altering or disabling genes can have major effects even on otherwise highly polygenic phenotypes; think dwarfism or mental retardation, for instance.)

Highlighting the fact that all humans share the same genes (even if their morphology may differ) ignores the fact that much of evolutionary change and biological difference is less about the development of novel proteins than it is about the regulation of those genes' expression—that is, the extent, the timing, and the location of when and where they are turned on and off. In fact, when the Human Genome Project first began, the number of human protein-coding genes was anticipated to be on the order of 100,000. After all, we are certainly more complex than *Zea mays* (corn) with its 32,000 genes, are we not?[17] As it turns out, we have a mere 20,000 genes (or fewer). So most human difference is driven by the turning on and off of those 20,000 genes in specific tissues at particular times. The same ones may be expressed in the brain and in the liver. They may get switched on by an attacking bacterium and switched off by a hot meal. Each one is like a hard-working parent who is multitasking for eight hours a day at the office and then rushing home at night to cook dinner, then attend PTA meetings.

The fact that we all share the same 20,000 genes does not rule out the possibility of phenotypic differences based on variation in the regulatory regions of the genome—promoters, enhancers, microRNAs, and other molecular switches. A better question than whether or not we have different proteins is whether or not we have different alleles. When we ask if there are alleles that one population has that are not seen in any other population—the parallel question to the unique genes inquiry—the answer turns out to be yes. As shown in figure 5.4, it is African populations that have the most "private" (i.e.,

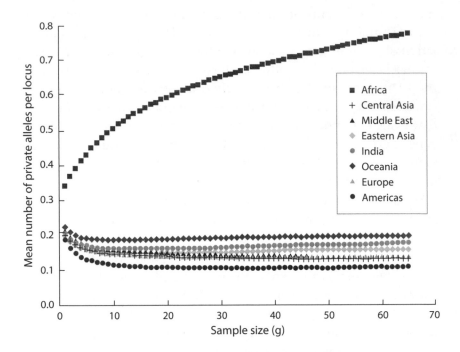

Figure 5.4. Private alleles (i.e., unique to that population) per locus with variation versus number of alleles sampled (in thousands) by ancestral group. (From Tishkoff et al. The genetic structure and history of Africans and African Americans. *Science* 324, no. 5930 [2009]: 1035–1044.

unshared) alleles. This is, of course, a reflection of the greater diversity in sub-Saharan Africa compared with those groups who suffered the population bottleneck in the migration from Africa to the rest of the world. But the point is that there is no a priori reason to rule out the potential impact of these private alleles in explaining group differences.

A third argument that the Left makes to discredit any genetic basis for observed group differences is that there has not been enough time—evolutionarily speaking—for meaningful differences to emerge. Stephen J. Gould is famously quoted as saying, "There's been no biological change in humans in 40,000 or 50,000 years. Everything we call culture and civilization we've built with the same body and brain."[18] According to this viewpoint, human evolution more or less ended with the emergence of anatomically modern humans in the Rift Valley. After all, sixty thousand years is but the blink of an eye compared with the entire history of hominids. And when we get to parsing differences between groups outside of Africa, that time span drops even more dramatically, as we saw in figure 5.2. However, crucial group differences can emerge not just through positive selection for novel mutations, but also through selection on traits that are highly polygenic, for which there is plenty of genetic variation already in the genome on which to selectively sort and reproduce. We already know that height and cognitive ability are highly polygenic, · influenced by thousands of small differences in the human genome. If those who are the smartest reproduce at higher rates than the less bright, an overall genetic shift in the IQ distribution could be achieved in a matter of a few generations (assuming that the reproductive and survival advantages of IQ were strong enough).[19] Sixty thousand years in this view is not a blink but an eternity.[20] So if there were differential fertility and survival premiums to different behavioral traits— not just IQ but also trust, grit, self-regulation, and so on—we could easily witness genetic divergence across the millennia.[21]

Indeed, this is exactly what some controversial scholars such as the anthropologists Gregory Cochran and the late Henry Harpending have argued in their book, *The 10,000 Year Explosion*. They posit that the Neolithic Revolution and rise of settled civilizations led to a condition in which human social arrangements—as opposed to the natural landscape—became the primary driver of changes in population genetics. The result, they argue, is that many differences today can be traced to the accelerated selective pressure that agrarian society introduced. Such pressure favors mental traits like advanced planning at the expense of physical endurance and other traits that would be more advantageous to hunter-gatherers. The time since the development of agriculture in a given society, they argue, is a good predictor of how the genetic landscapes of different populations have adapted to these altered demands for survival. Their case, while plausible, has not been made with the data at hand but instead represents a narrative that relies on circumstantial evidence. Although recent research suggests that evolution has not "stopped" due to technological and social progress; we do not know what forces drive recent selection or what effect they may be having in the contemporary world.[22] In other words, yes, humans are still evolving and genetically diverging from each other thanks to selective pressure (in addition to drift); however, claiming that the survival and reproductive gradients are different by continental and subcontinental locations—particularly with respect to social and mental skills—is unsupported by data.

The Left does not have a monopoly on substituting assertion for evidence when it comes to human evolution and group differences. The Right does a great job peddling its own untruths. For example, authors like Nicholas Wade, in his book *A Troublesome Inheritance*, focus on genotypes at one locus that display significant geographic or ethnic differences as a way to explain differences in group outcomes. It is not that a single gene cannot have a huge effect as *FOXP2* does, it's just that the ones that have demonstrated frequency differences by "racial" groups simply do not. For example, Wade and others often discuss the *MAO-A* copy number variant as the "warrior gene" because early candidate gene studies showed that this allele's presence predicted violent behavior. They then point out that the "violent" allele is found at higher frequencies in the black population. However, as we know from chapter 3, such candidate gene studies—and this one in particular—have not withstood replication tests that better control for population stratification. And even if they did, the allele explains a trivial amount of the variation in the measured outcomes, so it is hardly a solid foundation on which to build a genetic model of group differences in behavior.

A second mistake of the Right is to give too much credit to natural selection and too little to genetic drift.[23] The former, in which genetic variants provide a survival advantage, could theoretically lead to gene-based accounts of differences across groups. The latter, in which genetic variation is not tied to advantage but instead is a random process that "tags" groups based on geography and historical time, can differ by group (e.g., black versus white Americans), whose phenotypes like height or IQ can also differ but for environmental reasons that are spuriously correlated with the genetic differences.

We can see that purifying selection has occurred to accommodate the various environmental landscapes that humans have encountered as they fanned out across the globe. Very obvious examples include the prevalence of the sickle cell genotype—with its protective effect against malaria—which is present only in West and Central African populations, which have among the highest incidences of malaria in the world. (Other malarious regions have also developed genetically based resistance, such as *G6PD* deficiency.) Or the clear gradation in dermal melanin expression (skin tone) as predicted by distance from the equator and its intense sun exposure. Or even by body morphology, as evidenced by Allen's Rule, which suggests that in colder climates, warm-blooded organisms will tend to have shorter, stockier builds to preserve heat, whereas in hotter climes, big ears, noses, and limbs allow for better heat loss through a higher ratio of surface area to body mass. This relationship does indeed obtain in humans as well as in other endotherms.

The mistake that many genetic determinists make is assuming that because we can observe this clear relationship between environment and genetics in some physical characteristics, we can unproblematically expand it to highly complex human behaviors and mental characteristics. This is the problem of external validity, or generalizability from a small number of empirical observations. That we can see selective pressures at work in generating phenotypic differences in traits that rely on a small number of genes—such as skin tone, eye color, or lactose tolerance—does not easily translate to a clear relationship between a highly polygenic trait such as cognitive ability and the social or physical landscape. For example, when discussing body size, we can observe limb-length variation in human populations as predicted by Allen's Rule, but limb size is much less polygenic (it is largely controlled by a series of *HOX* genes) than is overall height.[24] And indeed, height fails to show the latitude-phenotype relationship as clearly. Compare Pygmies and Bantus (who occupy a similar relation to the equator) or Inuit and Swedes (who also live at more or less the same latitude). This does not mean that a highly polygenic trait cannot be subject to intense selective pressure; just ask chicken breeders who have quadrupled poultry weights in sixty years (figure 5.5).[25]

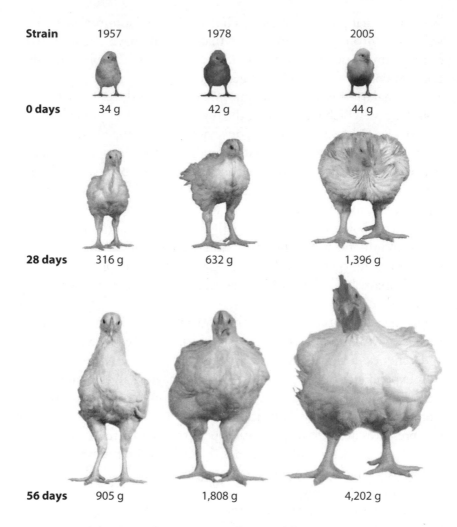

Strain	1957	1978	2005
0 days	34 g	42 g	44 g
28 days	316 g	632 g	1,396 g
56 days	905 g	1,808 g	4,202 g

Figure 5.5. The making of a monster. Chickens bred for size over six decades. Zuidhof, MJ, Schneider, BL, Carney, VL, Korver, DR, and Robinson, FE. (2014) Growth, efficiency, and yield of commercial broilers from 1957, 1978, and 2005. *Poultry Science* 93[12]: 2970–2982.

In addition to the complex and polygenetic structure of many behaviors at one point in time, rapid changes in economic fortunes during the last fifty years make simple genetic explanations of relative success by ethnic groups in the modern world all the more dubious. Although the last 10,000 years is certainly a plausible amount of time for geographic differences in the genetic architecture that shapes socioeconomic outcomes to emerge, 200 years is probably not, and 50 years is most definitely not. Yet we have seen Taiwan and South Korea, for example, go from poor societies with populations that could barely survive to countries with some of the highest living standards in the world.[26] Thus, there are probably better accounts than genetic ones for explaining geographic variation in standards of living and associated social outcomes, explanations such as institutional differences in the rule of law and so on. We turn to these in chapter 6. For now, research and theory suggest that genetic differences are a potential—but highly unlikely—explanation for national, racial, or ethnic differences in behavior and socioeconomic success, but such an explanation is a very difficult case to make. We can now see just how difficult.

CONTINENTAL ANCESTRY IS REAL, BUT WHAT DOES THAT MEAN?

Now that we know that there are significant genetic differences by continental origin (which correlate with but often belie racial categories), we can ask what it would take to answer an even more controversial question: do genetic differences by ancestral population subgroup explain observed differences in outcomes between self-identified race groups in the contemporary United States? That is, what if there are, in fact, distinct genotypes by race that matter for educational achievement and socioeconomic attainment over and above all the environmental differences that we also know matter? (We review nongenetic accounts of U.S. racial disparities in appendix 5.) Recall that although there are single variants that differ in frequency by continental ancestry, available evidence suggests such alleles cannot account for any sizable amount of the variation in IQ or socioeconomic status. But is it possible that all of the small differences in the genomes of black and white Americans could actually add up to explain a big part of the differences in cognitive or socioeconomic outcomes? And what methods are available to pool all of this variation?

We realize that this is a potentially "ugly" question with a long, often ideologically motivated history. There are solid arguments why scientists should not, in fact, even ask it. Namely, the damage done by such research exceeds the value of the knowledge produced by it.[27] But we raise it nonetheless to show how difficult it would be to actually answer it in a scientific—rather than ideological—way. While there are good methodological reasons for confining analysis of genetics to within a race, to write an entire book about social genetics and not address race directly would seem to imply we were copping out. And to write a U.S.-centric chapter about race/continental origin and not address the potential role of genes in explaining observed gaps in outcomes would make it seem like we were hiding something. Furthermore, such an elision would belie the spirit of this book: namely, that we should not be afraid to responsibly integrate genetics and social science. Finally, and most important, we address the controversial question of race, genetic ancestry, and achievement because we are fairly certain that with the large-scale integration of genetic data with social surveys, it will be addressed sooner rather than later by others—probably by scholars with an ideological, racist ax to grind. We would rather beat them to it to demonstrate how many assumptions need to be made and what kind of data are needed to make such an inquiry have any scientific legitimacy. So let us begin.

Perhaps an obvious way of proceeding is to attempt to directly examine whether all the small differences across the genomes of the average black and average white person in a data set "add up" in a way that suggests that one group has, on average, genetic signatures that predict higher levels of important phenotypes, such as educational attainment. There are at least two ways of "adding up" genomes. The first, which will be familiar to readers, is to use polygenic scores. The second, which we will discuss later, is the use of principal components.

What if we compared the polygenic scores for education between blacks and whites; would potential differences in the scores suggest that racial differences in schooling might be genetic? There are a host of challenges in conducting this comparison: some are technological and others are conceptual. The technological issues are presumably fixable, so let us start with those. First, the polygenic scores are constructed using samples of entirely (or nearly so) European ancestry. This presents several challenges including issues of external validity and also missing data. The question of external validity arises in part because the environments captured in the data to construct the polygenic scores do not include many environments in which nonwhites live (and get educated), so we should expect the polygenic scores to predict less well in black samples. Missing data arise because, as we discussed earlier in this chapter, African populations have much higher levels of "private alleles" (i.e., not shared with other populations). Since peoples of European ancestry do not have these alleles, the polygenic scores cannot be informative about these areas of the genome. (A related missing data issue is the use of imputation in genetic data; see appendix 6.) If there were a large sample of individuals of African ancestry (like the 500,000 person UK Biobank) with whole-genome sequencing, we could construct polygenic scores for education that could in principle overcome these two issues.

But even if such scores could be generated for each population, we would have a polygenic score made of apples for those of European descent and one composed of oranges (i.e., different alleles with different weights associated with them) for African-Americans. These respective polygenic scores could help us understand educational stratification within race, but they would be next to useless to explain cross-race individual or group differences. That is because any attempt to leverage the differences between these separately constructed scores would be violating the chopsticks rule: any allelic differences between the groups that were sizable should be picked up by principal components to factor out the potential environmental differences between the two groups.

The second approach we could take—measuring genetic ancestry directly and assessing its effect on outcomes of interest—also comes with serious drawbacks. At first blush it might seem promising. We already definitely know that there are distinct genetic signatures for what we would call continental clines, or what others might call race. Let us look at a map of the first two principal components (PCs) in two U.S. samples (figure 5.6), keeping in mind that what variation PCs pick up is highly dependent on the particular population sampled. (For example, in Europe the first two PCs map geographically North-South and East-West.[28]) Principal component analysis (PCA) is a statistical operation that calculates a single measure, the first principal component (PC1), which captures as much of the variation in the data as it can in a single dimension. With genetic data, PC1 is a way to capture variation in SNPs that tend to correlate across the genome and thus reflect ancestry. The second (and subsequent other) principal components extract the remaining variation that is unrelated to the first (or earlier) PCs—that is, some represent other dimensions of common ancestry.[29] If we know your score on these two PC measures, we can guess with reasonable accuracy (but by no means certainty) what boxes you will most likely check on the race and Hispanic origin questions on the U.S. Census form. Adding a few more PCs would improve our accuracy such that we might even be able to distinguish ethnic makeup within race groups. Other approaches that count ancestrally informative markers yield very similar results.

As we can see, if you scan from left to right horizontally along the *x*-axis (horizontal) in figure 5.6 (PC1), we see that blacks are spread out to the right while self-reported whites are clustered to the left.[30] Thus, scores on this axis can distinguish self-identified blacks from self-identified nonblacks pretty well. The *y*-axis (vertical)—PC2— further distinguishes the nonblack population into Asian (lower) and white (higher) identifiable groups. (Keep in mind that the directions— left or right, higher or lower—are completely arbitrary.) Meanwhile, Hispanics or Latinos are a highly admixed combination of these three groups, as shown in panel B of figure 5.6. Mexican-Americans show a higher proportion of Asian ancestry than other groups and Puerto Ricans the highest proportion of African, but European dominates this group.

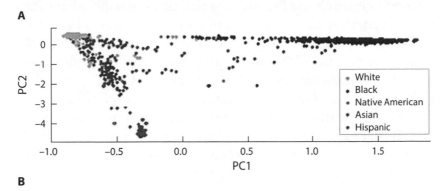

A

B

Add health (*N* = 2,281 individuals)

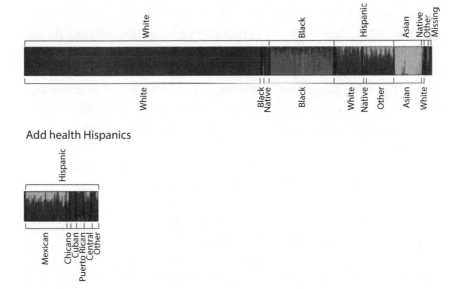

Figure 5.6. Admixture of U.S. respondents based on self-identified race in the National Longitudinal Survey of Adolescent to Adult Health. The bottom portion of panel B represents only those who reported Hispanic origin. (Panel B from G. Guo et al. Genetic bio-ancestry and social construction of racial classification in social surveys in the contemporary United States. *Demography* 51, no. 1 [2014]: 141–172.)

Conventionally, these PCs are used to factor out population structure (ancestry) in genetic analysis to avoid the chopsticks problem
mentioned earlier, which might lead to confounding environmental,
cultural, or historical differences with genetic ones as the root cause
of an outcome. As we described in chapter 3, in GWAS studies that
estimate millions of regressions that attempt to statistically associate
a single SNP with a phenotype, researchers often use ten or more
PCs in order to hold constant the rest of individuals' genomes. A
second typical use of PCs is to compare individuals across time and
geographic space in order to trace patterns of admixture between
populations as well as historical migration patterns.[31]

Yet a third way of using these metrics is still emerging and controversial: not merely as control variables but as the primary factors of
interest; that is, using these PC markers of "ancestry" to directly examine differences in socioeconomic outcomes. This type of analysis
has a methodological advantage: while the folk definition of race is
categorical (i.e., not along a continuum) and largely based on the
rule of hypodescent, ancestry is measured on a spectrum. Within our
categorical distinctions of white, black, Latino, Asian, Native American, and "other," there is a range and mix of ancestral genotypes, especially within a highly admixed society such as the United States.
Particularly among the black population, there is a distribution of
Africanness and Europeanness (and to a lesser extent Native Ameri-
canness). Indeed, the typical U.S. resident who identifies himself or
herself as black has, on average, 10 percent European ancestry.[32]

While it is not without precedent to use principal components as
a measure of African ancestry,[33] we proceed very cautiously. We dig
deeper in this direction for at least two reasons. First, this direction
of analysis presents a clear step that researchers will take at the intersection of genetics and social scientific questions, one that shifts away
from racial classifications toward continental ancestry (like similar
efforts in the medical and biological sciences). Second, we believe
our discussion can begin to describe pitfalls in this direction of inquiry and steer future efforts away from recasting the worst abuses of
past scientific approaches.

The use of PCs gives up on measuring the "correct," causal alleles for intelligence; instead, PCs can be deployed to see whether the percentage of African ancestry (or, conversely, the percentage European) predicts IQ or educational attainment. We can be agnostic about knowing which particular alleles matter for cognitive ability and instead ask whether an African American who is 20 percent European scores more or less on cognitive tests than a self-identified African American who is 4 percent European. (We could do the same in reverse for whites, but there is a lot less African ancestry in the white population due to the rule of hypodescent than there is the converse.)

A key dilemma in using this approach is that since we are using PCs as our main variable of interest, we are back to the chopsticks problem: how do we know that the genetic signature of ancestry, and not the history or geography that is associated with it, is what is predicting intelligence? The Deep South, for instance, still has the highest concentration of African Americans in the United States. Mississippi has the highest percentage black population of any state. In turn, Mississippi and its wider region have among the worst schools in the nation—for both blacks and whites. So maybe it is that we would be picking up an effect of being in Mississippi (assuming that the black population there is more purely of African descent than the black population elsewhere). Indeed, a recent paper shows that the percentage of European ancestry correlates with migration north and westward among the U.S. black population—that is, with better sociolegal climates for African Americans.[34] Or perhaps we would be picking up the effects of social class background, because those with more European heritage may come from more advantaged families. No matter how many things we try to measure and factor out, there could always be some unobserved environmental variable that clusters by PC.

But, there may be another approach to assessing the impacts of genetic variation on outcomes. Although PC values are largely shared within families,[35] the PC scores are not entirely identical within families.[36] In fact, in two data sets of largely white respondents, the sibling correlation is about .95 in PC1. For PC2 it is about .90.[37] In the

National Longitudinal Study of Adolescent to Adult Health, which has a representative sample of all ethnic and racial groups, PC1 has a sibling correlation of more than 98 percent in the whole sample (but noticeably lower for whites, at 89 percent). This means that there is some—albeit perhaps minuscule—variation in ancestry within families. You might recall from chapter 3 that by looking at genetic variation within families, we can slice the gene-environment Gordian knot. Siblings who are discordant in a given allele—for example, I have an A and my full-blooded sister has a G—got that way by random chance. Deploying sibling differences can not only solve the chopsticks problem for individual alleles, it can also assist in estimating the causal effect of polygenic scores or, for that matter, a PC—with a bunch of assumptions.[38]

The thought exercise is this: If we looked at differences between siblings in ancestry (or held parents' ancestry constant) and observed whether these within-family differences in ancestry (specifically the percentage of European ancestry) predicted cognitive ability, we may solve the chopsticks problem as well as skirt the causal allele identification/linkage problem.[39] (See appendix 6 for a lengthy discussion of linkage differences by race.)

The problems with this analysis would be twofold. First, we would need an enormous sample size. Since PCs are correlated between siblings at .98, the effective sample size for discovering effects using this approach needs to be more than 50 times larger than it would if we merely examined ancestry across randomly selected individuals in the population. Second, even if we had obtained such a large sample size that we could overcome the .98 correlation between siblings on their ancestry, and furthermore, we had found a result in our within-family models, what would that mean? Let us assume for the moment that we found that the percentage of African ancestry—that is, scores on the PCs—negatively predicted cognitive ability in Add Health. Such a result would raise the question: how? While it may or may not be true that brain development pathways could be implicated in test score differences, it is almost surely true that the percentage of African or European ancestry predicts physiognomy. That is, even within families, we are willing to wager that the sibling with more African genes is also the sibling with darker skin, curlier hair, and more West African facial features. There may even be other physical features that are less clearly racialized in the United States—such as height—that correlate with ancestry.

These physical features matter since they bring us back to square one. Namely, when Herrnstein and Murray or others make the argument that blacks are genetically inferior to whites with respect to cognitive ability, they imply a mechanism that is purely internal to the developmental logic of the fetus and the child. In other words, they argue that there are meaningful differences in genes that play a large role in central nervous system development independent of the observed range of environmental contexts. But our sibling difference methods—or any other approach so far developed—cannot separate the more context-independent (i.e., nonsocially mediated) biological effects from genetic effects that interact with the social system, such as when lighter skin is rewarded. That is, it could be that cognitive differences are genetically based, but the mechanism linking genes to IQ acts through social pathways (i.e., response to skin tone) rather than biological ones (i.e., brain structure). The darker-skinned sibling may get harassed by the police more often or get treated as less intelligent by his teachers (or parents for that matter) and this can, in turn, have real consequences for cognitive development. In other words, even if genes predict racial differences in IQ, they could do so because genes are good predictors of racial identification and treatment in society.

There has long been evidence—dating back to the days of W.E.B. Du Bois—that there is a pigmentocracy within U.S. black (and white and Latino) communities. More recent work has shown that this is not a uniquely American phenomenon but extends to Brazil, South Africa, and other nations with a creole, mixed population. We could try to measure skin tone and factor that out. But we cannot ultimately measure all the myriad cues about racial identity that we react to, especially since we may not even be aware of them. It could even be the case that African or European ancestry predicts height—as mentioned above—and that taller people are treated better in school, get more nutritional resources at home, and so on. Even though we do not generally think of height as a key dividing line for race, it does not mean that it is not silently associated—at the genotypic level—with the alleles that also differ by race.

For scientific accuracy, the reverse also needs to be said. We cannot conclude that the distribution of alleles does not explain racial differences in test scores through biological, nonsocial mechanisms. It could be that genetic variants that are not equally distributed by continental ancestry do, in fact, affect nervous system functionality in ways that we do not yet understand. It could even be the case that the very same genotypes that affect the amount of melanin in our skin or the shape of our noses also have other (i.e., pleiotropic) effects on our brains as well—independent of their effects through a discriminatory social system. Given what we know about history, our view is that a discrimination-based dynamic is much more likely to be the central explanation.

The near impossibility of a definitive, scientific approach as outlined above stands in stark contrast to the loose claims of pundits or scholars who assert that there is a genetic explanation for the black-white test score gap. In walking through the logic of genetic methods, we believe this discussion provides a cautionary tale for how scientists should proceed (or not) with investigations that combine questions of race, genetics, and socioeconomic or cognitive outcomes. With the outpouring of genetic data we are witnessing in society today, there will no doubt be further ventures in this direction. Clear efforts are therefore needed to sharpen the scientific questions that can be answered and also to guard against repeating past instances of pseudoscientific racism relying on ideologically motivated inferences from inadequate evidence.

That said, this dubious history does not mean that the consideration of genetics in racial analysis is always pernicious. The ability to control for genotype actually places the effects of social processes, like discrimination, in starker relief. That is, once you eliminate the claim that there are biological or genetic differences between populations by controlling them away, we can show more clearly the importance of environmental (non-genetic) processes such as structural racism. Controlling for genetic differences de-naturalizes the outcomes. For example, in an earlier study we conducted, birth weight differences among black identical twins predicted infant mortality more strongly than they did for white twins. Since this research design rules out genetic differences between the twins as well as behavioral differences of the mother (since the twins are affected in utero together by whatever she does or does not do), the results strongly suggest that racial disparities in neonatal health care post-birth are the most likely explanation of why low birth weight is more risky for blacks than for whites.[40] Or, if we were interested in the effect of colorism on health, we could compare full siblings who differ on skin tone while holding their genetic differences constant. If we then detect an effect of skin tone on, say, hypertension, this is much stronger evidence for the cause being the social stress of skin-tone based discrimination than if we merely make a comparison across individuals without controlling for genetic (and family background) differences associated with skin tone that might also affect blood pressure directly.[41]

This chapter has been a wild ride along an intellectual whitewater rapid. We have argued that common government definitions of race are flimsy social constructs inconsistent with actual genetic analysis. However, we also acknowledge that the notion of continental ancestry having distinct genetic signatures is a biological reality—albeit that biological "reality" is itself subject to social choices in how it is measured (namely, which populations are sampled to provide the template of genetic variation; see note 15 of this chapter). As genetic data become more available to the population, the mismatch between race and genetic ancestry (continental and subcontinental) should lead to a revision of racial discourse. That is, when many whites realize that they have African ancestry and many blacks discover their European ethnic origins through DNA testing, the one-drop rule might crumble and racial dichotomies could soften into more complicated nuances of admixture. On the other hand, as the sociologist Ann Morning has argued, "even with a familiarity with racial mixture that led us to put categories like 'quadroon' and 'octoroon' on nineteenth-century censuses, the one-drop rule hardly crumbled. In fact, it was reinforced in reaction to that awareness."[42] It may be that scientific knowledge has more power and authority to complicate matters than firsthand, intimate knowledge of racial mixing did. But it may not. Either way—as in the cases of marital sorting, class mobility, and fertility—social genomics reveals hidden dynamics of race that belie our intuitions. We cannot be afraid to look.

CHAPTER 6

THE WEALTH OF NATIONS

SOMETHING IN OUR GENES?

If genetics play a role within societies in helping to determine who prospers and who does not, who multiplies and who is barren, and even who pairs up with whom, then perhaps it can also help us understand why entire communities—nation-states, even—are rich or poor. In this chapter, we take some of the micro phenomena described in the previous chapters and scale them up to the macro level to tackle questions about the fortunes of entire nations. We explore how ideas and data from genetics and evolutionary sciences are shaping (and reshaping) interpretations of history. We focus on economic history: the population of nations and the geographic distribution of wealth in the world. This topic lies at the intersection of theories from two unlikely bedfellows: theories of historical economic development and those from population genetics. We discuss how the integration of these theories helps explain the pathway that we have taken and how we have arrived at a world in which some countries are hundreds of times richer per capita than others.

This enterprise—merging economic, anthropological, historical, political, and sociological analyses with ideas and theories from evolution, genetics, and biology—has been and continues to be controversial, with some entire disciplines condemning others. Anthropologists describe the work by some economists as "non-experts broadcasting bold claims on the basis of weak data and methods [that] can have profoundly detrimental social and political effects."[1] It is the academic version of a food fight. But at the same time, new discoveries are pushing us to consider novel directions of exploration to understand economic history.

A fundamental question in macroeconomics asks why some countries have thrived and others have stagnated over the past several hundred years. Data from the World Bank shows that nearly one in seven families live on less than $1.25 per day, although the distribution of these global poor varies widely around the world. For example, less than 1 percent of the populations in Europe live under this poverty line, while the corresponding fraction of people living on less than $1.25 per day in sub-Saharan Africa approaches half.[2] The average family makes a yearly income of $226 in Malawi but over $100,000 in Norway.[3] Noneconomic measures of well-being largely mirror these stark economic differences. The life expectancy at birth for women in Swaziland and Sierra Leone is less than 50 years, while it is more than 85 years for those born in Japan, France, and many other countries (the U.S. figure is 81).[4] How can we explain these extraordinary differences around the world, and what role—if any— might genetics play?

During the last few decades, explanations of country-level economic differences have shifted with the political winds. In the optimistic postwar period of the 1950s and 1960s, the Nobel Prize winner Robert Solow focused attention on technological innovation and capital accumulation—that is, whether countries invested in machinery, infrastructure, and the like. His model suggests that a major difference in the economic success of countries depends upon their choices to invest in the use of new technologies. Subsequent research broadened the idea of "capital" to add what economists call human capital— investments in education and skills—to the equation. Further extension included consideration of the research and development sectors in the economy. But throughout this period of refinement of the Solow growth model, the recommendations for development were relatively clear—countries should save and invest in technology and education. If they did, we would see rapid convergence in country success, as others mirrored the successes of the United States and Europe. The application of this prescription has met with mixed success. One possible reason for this is that the focus on the proximate causes of economic growth and development resulted in a one-size-fits-all set of rules that did not take into account more fundamental factors that may have been acting across much longer and expansive spans of time. These factors include the geophysical aspects of land, the institutions and culture of peoples, and, yes, even genetic variation across human populations.

ECONOMICS, MEET HISTORY; HISTORY, MEET ECONOMICS

A growing body of macroeconomic thought has begun to reach back in time to further understand the patterns of development we now see in the world, often combining historical, cultural, and sociological understandings of long-term processes. Economists of this school consider the long histories of institutions and cultural differences in thinking about the "deep" roots of economic development.

Many novel and controversial hypotheses have been proposed. As with the debates about nature versus nurture discussed in previous chapters, a key area of disagreement is what matters more—environmental factors or societal organizational structures. On the one hand, many believe that fundamental determinants of country-level income, development, and health are primarily *geohistorical*, a combination of relatively timeless geographical advantages (rivers, favorable climate, soil quality, disease burden, and so on) with favorable historical accidents (e.g., early domestication of animals). The confluence of resource advantages and technology "shocks" at the "right time." On the other hand, many others believe that key determinants of national success are primarily *institutional*—the formalization of property rights, the rule of law, democratic representation, and so on.

This distinction is important for many reasons, but a fundamental impetus for the interest is the desire to understand how to encourage, incentivize, and create development, success, health, and growth around the world. Broadly, two sides of the discussion are asking (and proselytizing about) whether successful development can be achieved by following a recipe or whether development comes "precooked," that is, predetermined. At one extreme, a focus on institutions as being fundamental suggests an ability to engineer success by following the examples of those who are currently successful. Think about development aid from the United States or the World Bank that comes with strings attached on what can be done with it. At the other extreme, a focus on geohistorical endowments and accidents suggests an inability to extrapolate from current wealthy countries—and is consistent with a lack of success stories of some developing countries modeling efforts on the United States or other rich nations.

As with most heated controversies, there are ample statistics and anecdotes to support each side. For example, some see the dramatic differences in levels of income and health between North and South Korea as the results of an "experiment" in the importance of institutions, with geography and populations held constant. Although these countries share a peninsula, a climate, access to trading partners, and a common ancestral (genetic) background, South Korea is one of the world's wealthiest nations,[5] boasting a per capita GDP of over $32,000 in 2012, whereas North Korea is one of the world's poorest countries, with a per capita GDP of $1,800 in 2012.[6] Many see the initial division and different alliances of South Korea (United States) and North Korea (Soviet Union) as further evidence of the preeminent position of institutions in the story of country development and success and the ability to "export" success. Implicitly, some assume that if the division were reversed, such that North Korea was overseen by the United States and South Korea had been overseen by the Soviets, it would be the North Koreans who would now be the envy of many countries, rich and successful, whereas South Korea would be the economic basket case.[7]

These views have been recently consolidated and explicated by the economists Daron Acemoğlu and James Robinson in their opus *Why Nations Fail*, in which they largely dismiss the environment as critical to economic success. Instead, the authors consider the development and expansion of pro-growth institutions (democracy, property rights, rule of law, and so on) as the set of factors that explain why some countries are rich and others are poor. Some countries find themselves with "extractive" institutions, through which small groups of individuals exploit the rest of the population (think about the ruling white minority in apartheid South Africa), whereas other countries develop "inclusive" institutions, through which many people are included in governance. Key to this argument are examples in which geography is held constant but institutions differ. Returning to the example of the two Koreas, the authors point to a recent history of inclusive governance in South Korea, and one of extractive, predatory rule in North Korea. It is not just about the benefactors (the United States versus the Soviet Union), but rather the specifics about the institutions that were formed in situ.

Think also about the economic performance of East Germany compared with that of West Germany or the economic growth trajectories in towns on the U.S. side of the Rio Grande compared with those on the Mexican side—similar environments, comparable populations, almost identical disease burdens, but different institutions and, thus, different growth trajectories. Such cases suggest that institutions matter a great deal in development outcomes across the world.

Others may concede that institutions matter in these micro comparisons, like between East and West Germany, but claim that the theory breaks down as we broaden the lens to ask about other comparisons: Why do the United States and Mexico, both democracies with reasonable infrastructures to support strong institutions, fare so differently? Critics of the institutionalist arguments also point to bordering states or counties within a country that experience different levels of economic development. The state of Kerala, for example, is underdeveloped compared with the rest of India. Northern Nigeria is the poor cousin of oil-rich Southern Nigeria. The seven emirates of the United Arab Emirates experience vastly different levels of wealth due to their distinct levels of natural resources, with Abu Dhabi displaying a per capita income of $50,000, whereas neighboring Sharjah enjoys only one-third of that amount. And even regions of the

United States marked by physical differences show enormous disparities: Appalachia versus the rest of the East.

Indeed, it seems oversimplified to suggest that environments do not matter at all. For example, tropical diseases reduce life expectancy in places like Zambia so that trained workers can expect to have only 10 years of economic productivity, whereas the corresponding figure for the United States is over 35 years. Agricultural grain yields from U.S. soil are about 10 times greater per acre than from African soils. And avoiding the threat posed by African river blindness means that Africa is the only continent where populations are concentrated *away* from the rivers and coasts, where the most fertile soil tends to be and where growth-stimulating trade is more easily facilitated.[8]

Current macroeconomic statistical analysis shows that approximately 50 percent (and as much as two-thirds) of differences in country incomes can be accounted for by a small number of geographic and environmental variables—measures of latitude, climate, and whether the country is landlocked or an island. About 13 percent can be accounted for just by knowing the latitude of the country alone.[9]

The geographic hypothesis does not stop with this handful of variables. It not only posits that current geographic conditions have direct effects on the current economic success of countries—such as hotter temperatures slowing down work and reducing productivity—it also argues that geographic conditions have indirect, historical effects on current success. For instance, hot conditions thousands of years ago may have increased the likelihood of the development of agricultural plots with lower productivity that in turn may further have reduced technological growth,[10] ending in a long-term reduction of economic growth that we see today. These dual processes mean that workers (and countries) in tropical climates nowadays face a double disadvantage from the environment: hot temperatures make it hard to work today because it is hot and because the land has not been well maintained due to higher temperatures for millenia. This cumulative process produces a rich-get-richer phenomenon—those countries that start off behind may never converge with those that started off ahead.

Jared Diamond (the author of *Guns, Germs, and Steel*) and others think that both effects are highly relevant—the initial geological conditions as well as the rich-get-richer effect of early success. First, geographic and environmental advantages such as a mild climate and lower disease burden were disproportionally available to populations in Europe and parts of Asia at the time of transition from hunter-gatherer societies to agricultural societies (i.e., the Neolithic Revolution about 10,000 BCE). The subsequent rich-getting-richer effect was the result of a variety of factors, such as the diversity of animals and plants available for domestication and the east-west orientation of Eurasia that facilitated the spread of innovations in agricultural practices (across a similar climate) that failed to occur in other landmasses, such as the African continent (with its north-south layout). Likewise, recession of the glaciers after the last ice age has led to a large variation in the quality of topsoil between and within some regions of the world.

According to the geography-is-destiny story, these initial advantages then further snowballed: the rich got richer, and then the richer got even richer still. Eurasian countries experienced a population explosion, and accumulation of technological innovations passed east to west and then back west to east along the Silk Road and Indian Ocean trade routes, as well as through migrations and invasions such as that of the Mongol Empire.[11] As early technological and population growth advantages accumulated over time, they allowed advantages in terms of military armaments, disease resistance, and technology (or "guns, germs, and steel").

These dynamic processes, which may take centuries or millennia to unfold, are called *deep determinants* of the success of some populations. Indeed, this theory suggests that even some proximate causes of success are rooted in prehistoric times. Examples of these factors include the claim that contemporary levels of development can be traced in part to whether a population's ancestors experienced a shift in their time preferences (trade-offs between present and future consumption) thanks to agriculture thousands of years ago.[12] Or whether a community farmed rice and thus developed a strong sense of community norms.[13] Or even whether plows were used to till soil, leading to a more gendered division of labor, compared with hoe-based farming societies, because of the upper-body strength required for guiding a plow through a field.[14] Very early on, geographic and technical advantages may have snowballed into current country successes and failures through a series of direct, and often quite indirect, effects.[15] The existence of these deep determinants of development create challenges for our ability to fully understand the key factors associated with economic development and to suggest policy remedies for underdeveloped countries.

Neither institutional nor geographic explanations can fully capture the successes and failures of countries in contemporary times. For the geohistorical critics, many of the factors are too broad and do not explain local and substantial differences in outcomes. Much of Europe enjoys the broad advantages of an east-west landmass alignment, yet England experienced the Industrial Revolution before the Continent did. (Though Britain did enjoy the physical advantage of coal deposits near the surface of the land.) As previously mentioned, North and South Korea and the former East and West Germanys are other examples in which geohistorical processes fail to fully explain the differences in outcomes.

The institutionalists also evince limitations in their explanations—notably their failure to explain why extractive institutions appear in some places but not in others. It is not very satisfying to point to the critical importance of specific types of institutions as essential for economic success without producing a set of predictive theories about why some countries develop extractive institutions while others develop inclusive institutions. In the extreme, the policy advice from an institutionalist to a poor country to "develop more inclusive institutions" may be as frustrating and unactionable as the advice from a geohistorian to "stop experiencing hurricanes and earthquakes."

The differences between the institutionalist and geophysical explanations of country successes and failures place rich countries, nonprofit institutions, and concerned world citizens in a bind. Many rich countries feel a moral obligation to help people escape poverty. Wealthy countries try to build infrastructure, support local industries and agriculture, expand education, build governance structures, and send cold hard cash and food. However, these efforts have been severely confounded by their mixed success as well as by vocal disagreements about the essential factors that determine economic development. Indeed, as the roles of geohistory and institutions are being sorted out, economists interested in the intersection of economic development, institutions, and geography have begun to explore another aspect of populations that might augment these grand theories: population genetics.

NEW CONTROVERSIES IN COUNTRY DEVELOPMENT AND GROWTH: ADDING GENETICS

A small group of social scientists has begun to put forth a new set of explosive ideas by introducing concepts from evolution, genetics, and biology into explanations of country-level success. These ideas fit well with the new focus of uncovering deep determinants of economic development. They incorporate linkages between environmental processes and the shaping of human societies by adding population genetics to the list of possible deep determinants. The research in this area is only beginning to emerge. Everything we discuss has been recently published and provides an immense opportunity for future inquiry, but this also means that the research is largely unsettled and in progress. The work has yet to coalesce around version 1.0.

Human population genetics measures the frequency of specific genetic variants and the historical and evolutionary processes shaping what we now observe across geographic and political borders. Combining population genetics with macroeconomics and economic history allows the concepts of natural selection, genetic drift, mutation, and gene flow between groups to begin being included as potential partial explanations for economic development, not to mention for other macro processes such as conflict and trade. One way that the geophysical environment can affect the economic success of populations is in shaping the genetic compositions of the populations themselves.

As we have already seen, the natural environment shapes genetic variation across countries and peoples. Exposure to malarial burden shapes our genetic architecture, leading to the presence and persistence of the sickle cell trait. Differences in exposure (or lack thereof) to sunlight shape the melanin content of skin. At the same time, random genetic drift over time creates myriad differences in the genotypes of populations—some important and others irrelevant. And these genetic differences are themselves mobile (called gene flow), as population genetics shapes environments and peoples expand, contract, migrate, and dominate, or are dominated by, neighbors and disease.

The genetic composition of populations may help determine which aspects of the environment are taken advantage of at a given time and place. The environments, far from being passive in the process, can further encourage development (e.g., the domestication of animals) or push out their potential human inhabitants (e.g., depopulation by epidemics). Some environments are cultivated by humans while others are not, and whether an environment is cultivated depends on its match with human genetics, which is also fluid. This process unfolds through a large-scale interplay between human genetic profiles and environments across historical time. But can this interplay help to explain the successes and failures of countries?

THE EMERGENCE OF (MACRO) GENOECONOMICS

As discussed in chapter 5, the natural history of population movements—which began in East Africa, with a trickle of migration out of Africa and throughout the (now) populated world—is relatively uncontroversial. To recap, African populations, as the original modern humans, hold nearly all of the human genetic diversity we see in the world. The smaller bands of humans who migrated out of Africa then took a portion of this genetic variation with them to populate the rest of the habitable planet. These basic genetic and historical facts can be seen in the profile of the world's genetic diversity, at least until the colonization of the world starting in the 1400s.

Populations in Africa had the most diversity, and the diversity trailed off as we got farther and farther (by foot) from East Africa, so that populations in South America had the least diversity (pre-1500).

But how have population migration events, bottlenecks, genetic drift, and other population genetics phenomena contributed to current developmental differences around the world? Are there aspects of genetic diversity that can help or hinder the health and economic success of nations?[16]

Quamrul Ashraf and Oded Galor published a paper in 2013 in one of the most respected economic journals, *The American Economic Review*, providing evidence that a "Goldilocks" level (not too low and not too high) of genetic diversity within countries might lead to higher incomes and better growth trajectories.[17] The authors observe that many countries with low genetic diversity (e.g., countries predominantly comprised of Native Americans, like present-day Bolivia) as well as populations with high genetic diversity (e.g., many sub-Saharan African countries) have experienced low economic growth, while many countries with intermediate genetic diversity (i.e., "just right"; European and Asian populations) have experienced high levels of development in the precolonial as well as the modern eras. Figure 6.1, which is from Ashraf and Galor's paper, plots country-level income per capita against a measure of country-level genetic diversity. The top of the "hump" is the Goldilocks level of genetic diversity, vis-à-vis economic success (i.e., the United States and much of Europe).

Some observers have noted how "convenient" it is that a genetic explanation of country-level differences in development would point to low progress in Africa and high progress in Europe and the United States. Indeed, the finding could be misinterpreted as showing that economic success, even at the country level, is "natural" and therefore immune to policy intervention.

The Ashraf-Galor theory also carries the implication that historical conquest by Europeans would be "good" for the development of the Native American populations by increasing their genetic diversity (if, of course, disease and genocide did not come along in the package). Although the authors have tried to discourage such oversimplified

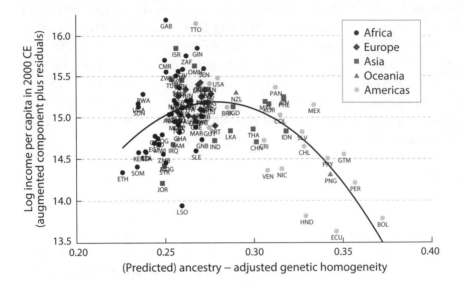

Figure 6.1. From Q. Ashraf and O. Galor. The "Out of Africa" hypothesis, human genetic diversity, and comparative economic development. *American Economic Review* 103, no. [1] 2013: 1–46.

interpretations of their findings and alert readers to alternative interpretations (including the likely important role of cultural traits), the nuances are often lost in favor of the more provocative explanations. These inflammatory interpretations and policy predictions suggest that we must exercise caution when considering this particular analysis if we are to avoid harmful missteps when integrating biology and genetics into economic and social explanations of development.

DIGGING DEEPER INTO THE THEORY AND FINDINGS

The basic theoretical idea (and hypothesis) is that populations and countries face a trade-off with respect to population-level genetic diversity. Borrowing measurements of diversity tied to concepts of race and ethnicity or economics (or income, or language, or religion), researchers have measured genetic diversity by appealing to the following thought exercise. Consider selecting, at random, two individuals from a population (i.e., a country) and asking what the probability is that they differ in their genetics at a particular site (or locus). Recalling that nucleotides (A, C, T, and G) form the 3-billion-long string of letters of a human genome, we can ask how much the two hypothetical individuals differ in their letters. Performing this type of analysis for a large number of people in each country will provide some idea of which countries have higher levels of genetic diversity and which have lower levels (figure 6.1).[18]

Ashraf and Galor hypothesize that countries face a trade-off between cooperation and innovation, which is as follows. Starting with a genetically homogeneous population, increases in genetic diversity are hypothesized to have two broad effects. With greater diversity come new ideas, culture, and practices that can be mixed and matched with preexisting ones to lead to innovation and creativity. Furthermore, the authors argue that higher diversity "enhances society's capability to integrate advanced and more efficient production methods, expanding the economy's production possibility frontier and conferring the benefits of improved productivity."[19] That is, diversity enhances specialization and comparative advantage—and thus growth—within a society. For example, if certain genotypes are good at detailed fine motor work and others are good at gross motor tasks, specialization can occur. The "fine motor people" produce what they are good at (i.e., "fine motor products"), and the "gross motor people" produce what they are good at making (i.e., "gross motor products"), and the two groups trade their goods. Economic theories related to specialization and trade suggest that the population is better off when the most efficient at each type of work dedicates herself to that labor task and then trades her products.[20]

However, there is a hypothesized downside to diversity. With greater genetic diversity comes mistrust and conflict. Ashraf and Galor argue that areas with genetic diversity that is "too high" have a greater probability of disarray, infighting, and conflict, which would reduce cooperation and disrupt the socioeconomic order, leading to

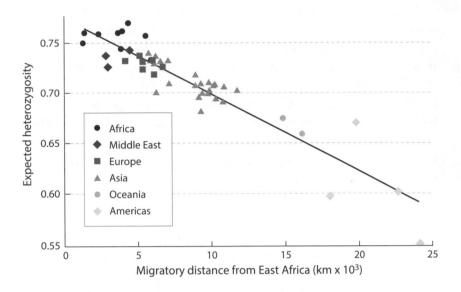

Figure 6.2. Expected heterozygosity and migratory distance from East Africa. From Q. Ashraf and O. Galor. The "Out of Africa" hypothesis, human genetic diversity, and comparative economic development. *American Economic Review* 103, no. [1] 2013: 1–46.

low productivity. So some countries (peoples) in Africa that have high levels of genetic diversity are predicted to face greater exposure to conflicts and mistrust of neighbors.[21]

These conflicting costs and benefits of genetic diversity lead to the proposition that a "middle level"—one that optimally trades off the pros and cons of genetic diversity—will lead to the highest growth and development patterns. Indeed, in an empirical analysis of more than 140 countries around the world, the authors find these trade-offs both in the past (around the year 1500) and the present (around the year 2000). They focus on the earlier time period to see whether the results are the same before and after the age of European impe-rialism. The authors can then examine how indigenous levels of ge-netic diversity among the conquered populations predict an impor-tant country-level outcome. If the results are similar, then the theory is strengthened. Findings from the early period also allow the authors to sidestep the claims of whether European conquests were "good" for the conquered country's long-term economic success.

What did Ashraf and Galor find? For countries with low genetic diversity (e.g., Bolivia), small increases in diversity (one percentage point) might have increased their population densities (a measure of economic development) by 58 percent by the year 1500[22]; countries with high genetic diversity (e.g., Kenya) that reduced genetic diversity by one percentage point might have seen an increase in population density of 23 percent by the year 1500.

Fast forwarding to present-day outcomes, the authors find that a one-percentage-point increase in genetic diversity would have raised a homogenous country's (Bolivia) income by 30 percent in the year 2000, and reduction of genetic diversity in an already diverse country (Kenya) by one percentage point would raise its income by 21 percent. Such effects are not trivial.

These two patterns for both the pre-Columbian period and the year 2000 produce the same type of results—a "hump-shaped" pattern, a Goldilocks effect—in which low and high levels of genetic diversity predict lower present-day country incomes but middle levels of diversity predict higher levels.

How do these patterns fit with more traditional explanations of differences in the wealth of nations? Ashraf and Galor propose that the genetic diversity findings are an addition to the story of economic development rather than an alternative to the geohistorical and institutional accounts. Indeed, through the use of statistical controls in their models, the authors take into account many mechanisms proposed in the literature, from the different times proposed for the origin of agriculture (Diamond) to the importance of institutions (Acemoğlu and Robinson). But they also suggest that their results have *additional* explanatory power—that the level of genetic diversity is a novel and important predictor of current differences in country-level outcomes. The finding that levels of genetic diversity may shape country development is important both theoretically and because it serves as an example of the new enterprise of leveraging theories, methods, and data from population genetics in an attempt to gain additional understanding of an important phenomenon.

But what do we make of the specific result? Are the successes and failures of countries in part dictated by long histories of genetic movements that have lasting effects on current outcomes? This is a hypothesis, guided by theory, with some initial empirical support. However, as Ashraf and Galor point out, there are several alternative interpretations of their results.

Even if we assume that the associations between genetic diversity and economic development show a cause-and-effect relationship,[23] the interpretation of the finding is far from obvious. Although the authors have their preferred interpretation, it is not possible to discard many alternative interpretations. For example, that genetic diversity is correlated with country-level racial and ethnic composition. The potential for these correlations means it is difficult to know whether the links between genetic diversity and country success are tied to Ashraf and Galor's theoretical ideas or are capturing other processes—such as histories of colonization, war, natural resource exploitation, and so on—that are tied up with race and racism.

Ashraf and Galor also acknowledge their inability to separate genetic diversity from the broader cultural processes that they are unable to measure in their data. This is significant and hinders our ability to make sense of the potential policy implications of the research. Genetic diversity could be "tagging" (i.e., statistically associated with) larger cultural processes related to how populations have distinguished themselves from their neighbors around the world; in this way, it is possible that measures of genetic diversity have very little to do with the varieties of As, Ts, Cs, and Gs. For example, cultural taboos against marrying immediate family members can create large differences in genetic diversity between populations.

A final issue with Ashraf and Galor's interpretation of their results is that they show only indirect and limited evidence that genetic diversity affects innovation or mistrust and conflict, which are their key proposed mechanisms. They show links between genetic diversity and scientific patents (i.e., their measure of innovation) but are unable to demonstrate these effects on a larger set of innovation outcomes. They also do not show evidence of increased mistrust, violence, and conflict. Without clearer evidence for the key pathways that compose their theory, the precise interpretation of their results is unclear. They may have found *something* novel about the predictors of economic development, but we do not know what it is.

How does their conclusion fit with institutional accounts? Are these findings at odds with the claim of Acemoğlu and Robinson that institutions are essential for development? Or do they provide the missing ingredient to the institutional account by using social genetics to explain the origins of institutional patterns? Perhaps the migratory distance from Africa is also related to differences in culture practices that have persisted to the present day, so that it is not genes per se (or their diversity) that affect current levels of country productivity, but rather that these measures of genetic diversity are "tagging" enduring cultural and institutional practices.

The claims of Ashraf and Galor's paper are extensive and pose an alarming hypothesis about the detrimental effects of genetic diversity on aggregate (i.e., country-level) production processes. As we explain, the hypothesis has not been demonstrated conclusively (an impossibility in a single study), so it is much too early for sweeping statements about the importance of the findings. Indeed, we may need to relearn the lessons of humility regarding scientific conclusions until more evidence is accumulated, especially given the legacy of scientific racism throughout the world. To further discern the mechanisms that link genetic diversity to economic development, new studies are needed at the various levels—micro, sub-country, and perhaps industry, factory, and production team—to provide additional evidence that these effects are detectable and "real" at units below the country level.

The additional claim in the study—of the beneficial effects of diversity in productive endeavors—has a long history, although expanding it to include the benefits of (genetic) diversity has been the subject of very few empirical tests. Here we may be at risk of extending (and overextending) the intuition from one domain (the gains in labor productivity resulting from novel ideas that complement one another) to another (that there are genetic sources of these complementarities). Even if there are findings that could begin to support the two ideas,[24] much more evidence is required to confirm the hypotheses. Like many claims in this nascent literature, these ideas could be true but as of yet are untested.

GENETICS AND WAR AND PEACE

Despite being thus far unsubstantiated, these initial but provocative findings have drawn other investigators into this field of inquiry. While not as well vetted as the Ashraf and Galor paper (yet), many new papers that extend these ideas to other outcomes have been published in quick succession. A paper by Enrico Spolaore and Romain Wacziarg examines whether population (country)-level genetic distance is related to the likelihood of conflicts between countries.[25] Their paper is also sweeping in scope. The authors examine interstate conflicts and wars between 1816 and 2001 for more than 175 countries and ask whether countries that are less similar in their genetic makeup (greater "genetic distance" from one another) are more likely to engage in wars. Think of wars over long-standing ethnic differences.

However, the integration of genetics with international relations turns old results inside out. Instead of dissimilar groups being more likely to have conflicts, the new findings suggest that (genetically) similar groups are more likely to have a conflict or war with one another. This discovery is not as strange as it might first appear and could have many explanations that are not necessarily "genetic." First, genetically similar populations are likely to live next door to one another, potentially generating wars and conflicts because their (genetic) neighbors are "getting on their nerves." In order to address this issue, the researchers statistically adjusted the analysis for geographic distance between countries, and the genetic distance still matters. Another "genetic" phenomenon that has a strong "social" implication is genetic admixture. One way for populations to be genetically similar to one another is through a (potentially forced) admixture of genetic pools following historical conquest, trade, and similar processes—think the War of 1812, where the two battling countries (the United Kingdom and the United States) have a small genetic distance and a reason for the conflict stems in part from a previous conquest. Then we would not be as surprised if genetically similar countries were likely to have future conflicts, because the similarity was partially generated through past conflicts (conquests and admixture). Again, the researchers make statistical adjustments for these (and other) predictors of conflict and war but still find a remaining effect for genetic similarity for the populations. Like the Ashraf and Galor findings linking genetic diversity and economic development, these results linking genetic distance to interstate conflict are new and not fully understood.

NATURAL SELECTION, MUTATIONS, HEALTH, AND PERSISTENCE

Other economists are beginning to consider the role of population genetics and its interaction with environmental resources through historical time. These researchers are interested in how certain aspects of population genetics may affect differences in growth patterns across countries, though instead of positing "optimal" levels of genetic diversity, they delve into specific sequences of our genome to ask whether distinct changes in the genome over time may have enabled different populations to take advantage of their environment. A case in point is the potential variation in economic success that may be attributed to a truly localized change in our DNA—the lactase gene.[26]

The economist Justin Cook has shown that the (genetic) ability to digest milk after weaning, which appeared early in human history, conferred large advantages in population density around 1500 CE (a 10 percentage point increase in the prevalence of the beneficial genetic variant in a population was associated with about a 15 percent increase in population density).[27] Since other studies have shown that differences in economic development in the past have been remarkably persistent, the implication is that (relatively) small changes in the genome at the right time and in the right place (during the

Neolithic Revolution in areas able to raise cattle) can lead to large, persistent, accumulating differences in economic development across countries.

Macroeconomists have also begun exploring how population genetics might affect economic development through the health of populations, which subsequently affects productivity and income. For example, Cook has shown that populations with immune systems that are "genetically diverse" had a health advantage in the premodern period.[28] The idea is that pathogens evolve to target specific weaknesses in immune function, and populations with limited genetic diversity (and hence limited diversity in immune response) are at higher risk of infectious pathogens spreading and reducing the health of wide swaths of the population. However, genetic diversity within the population can protect against such widespread health insults and constrain the epidemic spread of illness.

For some social scientists, this hypothesis is quite intuitive, because it follows from well-established results of mathematical game theory in economics. Consider the population game called Matching Pennies, in which each player turns a coin to show either heads or tails. In the case of a simple two-player version of the game, player 1 wins if both pennies are the same (heads/heads or tails/tails), and player 2 wins if the pennies differ (heads/tails or tails/heads). If participating in this simple game, what should a player do? What is the *optimal strategy*? First, let us ask whether there is an "easy" answer—whether player 1 should always choose heads or always choose tails, regardless of what she expects player 2 to choose. Unlike many important examples from game theory, Matching Pennies does not have an "easy" answer—a so-called pure strategy that dominates all other strategies in which a player would always choose heads or always choose tails. Indeed, a players' optimal strategy is to pursue a "mixed" strategy, sometimes playing heads and other times playing tails, because each player does not want her opponent to be able to guess her choice. Therefore, the optimal strategy to Matching Pennies is to randomize your choice. This simple game has many real-world implications. In football, a simplification of the game is to think of each team (offense and defense) as deciding on whether to run or pass (and defensively, whether to prep for a pass or run). If the defense and offense choose the same option, the defense is more likely to win, but if they choose different plays, the offense will more likely win. And you do not want your opponent to know your play. So game theory tells us to partially randomize strategies—to sometimes take a strike at the end zone from midfield on first down or to throw a screen pass on third and long.

Why does this detour into game theory matter for population genetics and health? Consider a similar game played by humans and pathogens. Humans create a set of defenses to pathogens but cannot defend against all possible attacks. On the other side, pathogens "choose" a specific route for infecting humans. The game is dynamic: once humans are successful in fending off a specific attack style, the pathogens evolve to attack in a different way.[29] It's like Matching Pennies (and run versus pass) in that if both humans and pathogens choose the same route, humans win by having a defense in place to stop the play (the attack). If humans and pathogens choose different plays, pathogens succeed in infecting the humans by getting around their defenses—by faking a run and throwing downfield. Thus the game has a clear prediction: human populations who are able to follow the mixed-strategy paradigm will be most able to survive and thrive when faced with pathogens, both old and new.

Indeed, Cook finds something seemingly improbable—genetic variants again seem to play an important role in country-level outcomes. Specifically, increases in population-level immune genetic diversity (through the human leukocyte antigen [HLA] system) lead to increases in population (country-level) life expectancies.[30] But maybe he is just illustrating a relationship that has already been established. Perhaps HLA diversity is related to country-level migration inflows (admixture would increase HLA diversity), and he is "finding" that countries with inflows of migrants live longer than countries with fewer migrants. To further clarify his hypothesis, he proposes an additional test. He notes that during the second half of the twentieth century, several vaccinations were developed to combat infectious disease. These vaccinations actually remove the advantage of having HLA diversity in a population. That is, modern science and medicine is substituting for "natural" (genetic) defenses against illnesses—at the population level. In earlier times (in disease-rich environments with lack of medications), genetic variation acted as a buffer against disease, leading to country-level differences in life expectancy based in part on genetic differences. But now that the environment has changed, with new medications and vaccinations, the previous genetic advantages have been largely eliminated. He tests this implication in his data, and his results confirm the idea. He shows that while there were large advantages in life expectancy in the mid-twentieth century for countries with high HLA diversity, these advantages of HLA diversity disappeared during the latter half of the twentieth century as vaccines and new medication proliferated.

This initial advantage of HLA diversity followed by the elimination of the advantage as the environment (i.e., access to medicine and vaccines) changed is another example of the interplay between genetics and environment, although it occurs at the population level over historical time. We saw a similar story unfold in the earlier example of the lactase gene, which confers advantages only in environments that have the ability to foster the domestication of animals. With no cows, goats, or other domesticable mammals, the gene provides no population advantage. We address the interplay between genetic and environmental factors in chapter 7.

Where does this new evidence leave us in our quest to understand the determinants of the wealth of nations? While not overturning geohistorical and institutional approaches, population genetics concepts and data do seem to contribute to the overall story. Like genetic analysis more generally, the approaches combine both top-down and bottom-up perspectives. Similar to the candidate gene (bottom-up) approaches discussed in previous chapters, some macroeconomists are focusing on specific genetic variants, like the lactase gene, with a clear, hypothesis-driven, mechanistic lens. It is incredible that a single mutation could be responsible for such massive differences in country-level success across hundreds of years, and this possibility suggests opportunities for expanding geohistorical perspectives on macroeconomic growth patterns to reckon with the influence of population genetics. Similar to GWAS (top-down) approaches, there are also efforts by economists to uncover associations between broad genetic measures (e.g., genetic diversity and/or genetic distance) and economic development, where a mechanistic understanding is far removed from the exercise. Additional research will probably add to our understanding of the key predictors of development and migration. But as new results emerge, they will likely increase the danger of "naturalizing" (taking as a given because it is in the genome) the wealth of nations. It is an impulse, we think, that should be combatted because (as we will see in chapter 7) for many human behavioral outcomes, it is not genes *or* environment that best explains reality, but rather genes *times* environment.

CHAPTER 7

THE ENVIRONMENT STRIKES BACK

THE PROMISE AND PERILS OF PERSONALIZED POLICY

Most science examines the typical (or average) effect of an environmental factor, or a policy (or a gene, for that matter), on a given outcome. This is the bread and butter of randomized controlled trials, which contrast the average outcome for people who received the red pill compared with the average outcome for people who received the blue pill (a placebo). Focusing on the average effects makes sense if we are asking whether the red pill typically outperforms the blue pill (or whether access to Medicaid increases healthy outcomes compared with no access to Medicaid, or whether lead exposure reduces IQ compared with no lead exposure). Of course, extreme outcomes may matter as well. If a treatment has an average effect that is positive but has a very severe negative side effect in a small number of people, we may want to prevent it from going to market.

Imagine you are an intrepid social scientist who has devised a novel way to teach reading to first-graders. Being the careful scholar you are, you implement a randomized controlled trial in which some classrooms get your approach to literary education and others get the traditional teaching methods. You measure reading ability at the beginning of the school year and again at the end. The following summer you eagerly analyze your data and find that there was a statistically significant 15 percent (one standard deviation) better performance on reading tests in your treatment group than in the control classrooms. You jump for joy. That is a huge effect in the education literature. You will surely be able to publish your study. And more important, if you can convince the educational bureaucracy, perhaps you will be able to implement your curriculum system-wide and help lots of kids. But then you look more closely at the data and find that while the average difference between the two groups was that one standard deviation (15 percent), the individual-level effects were all over the map. It turns out that a few kids in the treatment group registered enormous gains: they more than doubled their reading scores. In fact, their drastic improvement (which was not matched by anyone in the business-as-usual control group) accounted for the entire 15 percent effect when averaged across all of the treated kids. In fact, another handful of kids in your treatment group actually got worse in reading, though by a small amount. And in the control group, there were no instances of backsliding.

Clearly, you are suffering from a case of highly heterogeneous treatment effects. For the typical or median kid, your program makes not a whit of difference; for a few it is even damaging. But for another group it is like a magic elixir. If you could only figure out which kids respond well and which kids do not. You are not alone; it turns out that such heterogeneous responses to social interventions are more the norm than the exception. Sometimes the differential responses fall along an identifiable dimension, such as income level or gender or even body mass index. But more often, the font of variation remains obscured in the shadow of the average effect. If only we had a prism that could refract the white light of average treatment effects into the rainbow of heterogeneous ones, we would be set. That is the motivation—the holy grail if you will—behind gene-by-environment research by social scientists.

The more we understand about the interplay between environmental and genetic factors, the more we understand that we miss a lot of the story when we focus only on the average effect of carrying a genetic variant or being exposed to a specific environment. Since genetic factors can filter and refract how the environment affects us, it could be that the red pill works great for people with gene variant A but poorly for people with gene variant B. But at the same time, the red pill could be worse than the blue pill for people with gene variant B, leading to an overall average treatment effect of zero absent genotype information. Knowing more about the interaction between genetic and environmental factors might show us that we should always give the red pill to gene-A types and the blue pill to gene-B types to obtain the best outcomes. The fields of pharmacogenomics and personalized medicine are asking these exact questions, but the idea does not end at medications. It could be that some policies help many people but hurt (or are irrelevant for) other people. If so, then perhaps we need to develop fields specializing in personalized policy and "policy-evaluation-omics."

Indeed, studies documenting the contingencies between genes and environments are leading researchers to question and reinterpret the received wisdom both from genetics and from social science and policy evaluation. These contingencies also suggest that, as we move away from a narrow focus on the average impacts of pills and policies, a combination of experts across the biological and social science disciplines working with one another will need to provide us with new and different forms of evidence, including a focus on the distribution of effects (rather than just the average) and measurement of the genotypes of subjects in order to inform randomized control trials (which is beginning to happen).

GENE-ENVIRONMENT INTERACTION AND THE CASE OF IQ

Some gene-environment interactions are sort of obvious. Genes that help store fat from caloric intake might be extremely beneficial in hunter-gatherer environments, where food sources are sporadic and uncertain, but these same genetic variants, performing the same biological functions, might instead lead to morbid obesity or metabolic syndrome in modern food environments rife with cheap and dense calories. The genes are doing the same thing in each case, but the environment is shaping the eventual phenotypic outcome of the relevant genetic variation. Likewise, a genotype that enhances risk-taking behaviors in humans could have been advantageous when we had to hunt and kill our dinners, but in today's society it might be more predictive of incarceration. Again, the environment is not necessarily shaping the internal, biological function of a gene (although that may also be possible); it is instead shaping the downstream outcome.[1] These examples should force us to reckon with the shortcomings of ignoring the environment while trying to ascertain the influence of genes on behavior. On one hand, an exclusive focus on genotype is helpful if we are examining very basic biological functions that are not socially or environmentally mediated, such as the coding of a protein that has the letters CGC (rather than CCC) in a sequence at a particular locus thus leading to a change in insulin production. On the other hand, for outcomes that also have social determinants, these examples suggest that a narrow emphasis on genes alone may be untenable and that gene-environment interactions are the rule rather than the exception.[2]

A clear theme in the gene-environment literature is that it is not so much nature versus nurture but rather the interrelationship between nature and nurture that we should be investigating. In hindsight, perhaps, this is an obvious point. It is not that gene X does Y, but rather that gene X, shaped through exposures to environment A, might do Y and, if alternatively exposed to environment B, might do W. The interplay works in both directions. The effects of being exposed to environment A may differ based on whether a person carries an A or a T at a certain position in the genome. These differences are often presented by tracing out the norm of reaction of a genotype across a range of environments. Figure 7.1 shows the hypothetical phenotypic outcome for genotype A across a range of environments, as well as the outcomes for genotype B. Unless the curves run in parallel, the different norms of reaction suggest gene-environment interplay. Regardless of how we draw it or name it, the idea is that the impact of genotype on behavior is contingent upon the environment

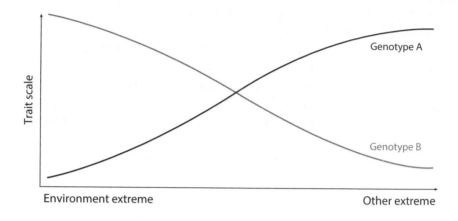

Figure 7.1. A cross-over gene-by-environment interaction effect might look like this.

and the impacts of environments on behavior are often contingent upon genetics. And, to add further complexity, we actively choose and shape our environments based partially on our genotypes (i.e., carriers of A variants at a locus may seek out different environmental niches than carriers of T variants).

What about the role of gene-environment interactions in socio-economic outcomes? Early research on the heritability of IQ and economic status pointed to an intriguing finding that bears mentioning. Recall from chapter 2 that certain scholars have argued that we should actually strive for a world in which socioeconomic measures of success are wholly genetically determined—that is, in which heritability is as close to 100 percent as possible. Others would view 100 percent heritability as a dystopian nightmare. As proponents see it, any social effects—especially those of family background—are inefficient and unfair. The empirical reality (at least one version of reality based on earlier twin studies) is that for some groups this genetic Shangri-La has indeed been more or less achieved, whereas for others it is still a distant dream.

The psychologist Eric Turkheimer (see figure 7.2) has argued that there is little equality when it comes to the effect of genes.[3] He ob-

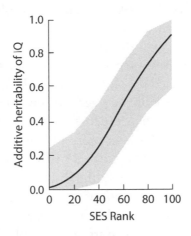

Figure 7.2. The heritability of IQ by socioeconomic status (SES) (From E. Turkheimer et al. Socioeconomic status modifies heritability of IQ in young children. *Psychological Science* 14, no. 6 [2003]: 623–628.)

served that genetics seemed to matter less in predicting the intelligence of twins coming from the low end of the socioeconomic distribution compared with that of twin pairs born to richer parents. That is, the gap between the similarity in identical and fraternal twins—which is the measure of genetic effects or heritability in twin models—was larger the higher up the economic ladder one went. According to Turkheimer's interpretation of these results, those born to families who are socially advantaged receive the financial and nonfinancial resources to insure that they reach their full genetic potential—at least when it comes to cognitive measures. Social disadvantage eliminates the genetic advantages that would have otherwise "naturally" emerged in an equal-opportunity society, thus it squelches innate differences in ability, leading to a dire situation at the bottom where genes do not matter.

This theory makes intuitive sense. Think of a discriminatory social environment in which everyone who is black is unable to attend college or apply for a good job. Such a dynamic with respect to race was observed as far back as 1967, when the sociologists Otis Dudley Duncan and Peter Blau coined the term "perverse equality" to describe a situation in which the class background of African Americans had relatively little influence on their occupational attainment.[4] Discrimination repressed the children of black doctors as much as it did the offspring of black ditchdiggers. Meanwhile, tokenism insured that for each generation there emerged a "talented tenth" (to use the language of W.E.B. Du Bois)—a cadre of black professionals whose outcomes could not be well predicted by their backgrounds.

Although Turkheimer was talking about IQ scores and Blau and Duncan were discussing occupational outcomes, their stories coalesce nicely.[5] It is a story that can straddle the middle ground between the nature and nurture camps. In contrast to the blank slate—pure nurturance theories that are often fashionable with the academic Left—Turkheimer argues that genetics do matter—and yet his story acknowledges the critical role of unequal environments. Intuitively, his idea suggests that a redistribution of resources would not only serve to equalize outcomes (i.e., fill in the hole in which low-socioeconomic-status [SES] kids find themselves), it would also lead to greater economic efficiency by unleashing the underutilized, latent genetic talent of individuals trapped in low-income environments. In this way it may be cost neutral or even revenue positive from a public economics standpoint.[6]

But are such claims of a gene-environment interaction accurate? Perhaps instead of an increased effect of environmental differences within the lower end of the socioeconomic distribution, it is actually a weaker genetic effect. That is, what if poor families tended to have less genetic variation? That would make fraternal twins seem more like identical twins by virtue of being more than 50 percent related and would generate the same pattern of results that Turkheimer ascribed to environmental factors. Such a situation could arise due to a greater degree of assortative mating in low-SES groups.[7] The observed pattern of results would be the same, but the implications would be entirely different and might suggest that the intervention should be in the *mating market* rather than the *educational system*.

With twin models, it is hard to know exactly what is going on since the genotype is inferred rather than actually measured. That is, we are assuming that identical twins are more genetically similar than fraternal twins, which seems logical. The question is, how much more similar are they? Specifically, how genetically alike are fraternal twins, and does the answer to this question vary by SES? For example, it could be that thanks to a greater degree of genetic sorting (positive assortative mating) in the low-SES group, low-SES fraternal twins are more related than 50 percent on average. The closer relationship would lead to an underestimation of heritability (genetic effects) in this group. Alternatively, it could be that, as Turkheimer posits, genetic sorting is the same for both high- and low-SES families and genetic effects (heritability) would be the same if the environments were equalized between the two groups. There is no way to distinguish between these two plausible explanations (which, in turn, carry very different implications for policy) without measuring the genome (and/or the environment) directly, because with twin models we do not actually know the degree of underlying genetic variation (genetic assortative mating) within subpopulations.[8]

We reasoned that the next step would be to measure the previously unmeasured genetic predictors of academic achievement and ascertain whether the genetic predictors more accurately predict academic achievement in higher-SES households than they do in low-SES households. In order to measure this variable—the underlying genetic architecture of academic achievement—we needed to create a new variable in our data to measure the genetic potential for educational attainment.

To create this variable, researchers used data on the genotypes and educational attainments of over 100,000 people in order to conduct a GWAS analysis that could measure the total additive effects of genetic variants in predicting educational attainment. As discussed in previous chapters, a GWAS analysis combs the genomes of the sample participants to discover which genetic variants show up disproportionately in those people with the outcome of interest—in our case, highly educated people. Combining all the variants that predict educational attainment then creates a polygenic score. The polygenic score can then be recreated in any other data set that also has genome-wide data and schooling (or other) outcomes to conduct additional analyses to uncover potential mechanisms that underlie the polygenic score's relationship with educational success (as long as the new dataset is drawn from the same basic population in terms of broadly defined ancestry). The results from the GWAS were a promising first step. The score could account for about 2–3 percent of the variance in educational attainments across tens of thousands of people in the data. Heritability estimates for educational attainment using twin studies suggest a figure near 40 percent, so the initial GWAS findings are subject to the missing heritability question we outlined in chapter 3. But it is a start (and, as mentioned earlier, has since been improved upon).

So with the polygenic score for education that was constructed for two independent samples, we attempted to test the Turkheimer argument (and the alternative hypothesis that socioeconomic status does not affect heritability) by asking two related questions. First, did the distribution of the polygenic genetic score (i.e., how much variation there was among respondents in their scores) differ by social class background (as measured by maternal and/or paternal education)? This would be a way to assess the possibility that it was the genetic landscape, and not necessarily the environmental landscape, that varied by socioeconomic status. Second, did the actual impact of the education polygenic score vary by class background, for whatever reason?

The answer to the first question was "no," the spread of scores did not vary by social class. In two data sets—the Framingham Heart Study and the Minnesota Twin Family Study—the standard deviations of the raw scores were the same for children of mothers (or fathers) who had only a high school education (or less) compared with those respondents whose parents had at least some additional schooling (other measures of the distribution also did not show significant differences).[9] So far, this was good news for Turkheimer's interpretation of an environmental explanation for the interaction effect. The results were consistent with the view that the muted genetic effects for poor children were not evinced because of differences in genes but might be different because of differences in environments.

However, we then tested whether the effect of the offspring score varied by parental education, and it did not. Though an imperfect test of the argument,[10] the results suggested that genetic advantage or disadvantage had the same magnitude of effect whether a child was from a high- or a low-SES family—contrary to what Turkheimer found in the twin studies.[11]

Although the polygenic score approach enjoys the advantage of direct measurement, it suffers from the significant disadvantage of capturing only a small portion of the total additive genetic effect.[12] It may also suffer from a deeper issue. Because the score was calculated from a meta-analysis of cohorts across a wide range of environments, it may be capturing precisely the genetic effects *least* influenced by the environment. That is, the genetic effects that appear in the polygenic score are those that are relatively similar across the many data sets (from many environments and over different time periods) in the discovery analysis. For example, the initial education polygenic score was created using 54 data sets. When we then use the polygenic score to test whether the genetic effects interact with the environment, we may be setting ourselves up to fail the test. Because our genetic measure is created by averaging results across a wide range of environments, we are likely using a genetic signal that is relatively robust to (unaffected by) environmental differences.[13] That is, we might think that the variants that operate across all the environments might be exactly the ones that do not exhibit much gene-environment interplay. However, what if in social democratic countries, like Sweden or Norway, a certain allele that predisposed individuals to be more cooperative and less competitive led to significantly greater educational success, but in more competitive, laissez-faire capitalist settings, like the United States or Australia, that very same allele had a negative effect on educational performance as a result of different cultural norms or expectations? In both societies this allele would be a predictor of education, but in the pooled analysis, its effect would be zero because Scandinavia would cancel out the United States. But this is exactly the kind of gene-environment effect that we are looking for.

An alternative approach to incorporating polygenic scores into gene-environment research is to create a different type of polygenic score measure. Rather than use a polygenic score that best predicts the average level of an outcome, like education, we may want to create a polygenic score that best predicts environmental sensitivity for use in our gene-environment analysis. Although this approach is still being devised, a first step is an approach that uses GWAS analysis to predict variation in a phenotype rather than the level of a phenotype; this approach is sometimes called vGWAS and is an area of active research.[14]

So although the issue of what is going on across class lines is far from resolved, at the very least these newer results that use polygenic scores throw into sharp relief the potential foolishness of assuming that when we observe a difference in genetic effects by some population segment—race or class or geography or family type—this reflects a "true" gene-environment interplay.

As if understanding the interplay of genetics and environments was not complicated enough, these two factors seem to interact with one another in a very complex and dynamic feedback loop. This loop makes disentangling cause and effect all the more challenging.[15] After all, genotypes can select into environments as long as humans (especially parents) enjoy some modicum of choice regarding where they live, with whom, and how. Separating gene-environment interactions from gene-gene or even environment-environment interactions is no easy feat, but it is a critically important one in order to understand the roots of human behavior and make the right diagnoses for policy.

Turkheimer was a pioneer in positing gene-environment interactions with his twin analyses stratified by SES. But it was not until almost twenty years later that the first molecularly measured gene-environment interaction paper was published. The researchers who first pinpointed particular genes and claimed that their effects varied by environmental context did so by deploying a unique survey from New Zealand that followed for several decades a birth cohort from Dunedin, on that country's South Island. The investigators came from a range of disciplines. Terrie Moffitt had trained as a clinical psychologist and Avshalom Caspi as a developmental psychologist.

They teamed up with other psychologists and with geneticists to examine why some children exposed to violence and maltreatment during childhood were less adversely affected than others. The question of resilience was an important and well-researched area in social science but until this study had not used genetic data. Here the researchers examined variation within a single gene, *MAO-A*, among a large sample of male children exposed to different levels of maltreatment. They found, unsurprisingly, that children exposed to violence were more likely to be antisocial and violent adults. The more intriguing result was that the links between maltreatment and adult misbehaviors were shaped by which genetic variant the boys carried. Maltreatment of males with a "low activity" *MAO-A* genotype (i.e., one that tended to produce a lower amount of the protein) had more than twice the impact for determining adult antisocial behavior as maltreatment of males with a "high activity" genotype.

MAO-A codes for an enzyme that degrades neurotransmitters, like serotonin and dopamine. Indeed, the *MAO-A* gene had been labeled in the early social genetics literature as the "warrior" gene because of its putative links with aggression (see chapter 3).[16]

There was a lot to like about this study: It was both pathbreaking and carefully undertaken. The measurements of maltreatment during childhood were based on well-validated instruments to assess this sensitive subject.[17] The candidate gene of interest was carefully chosen on the basis of previous scientific work, work that experimentally showed effects of the gene using animal (mouse) models.

The results of this study, along with a companion paper examining a similar question but different gene (*5-HTT*), appeared in the journal *Science* in 2002 and 2003, respectively.[18] *5-HTT*, a serotonin transporter gene involved in the reuptake of serotonin at brain synapses, is a target for antidepressant medication (SSRIs). The gene has been studied in experimental animal (mouse and monkey) models that showed differences in the effects of stressful environments on the level of stress response with different variants of the gene. For their *5-HTT* analysis, Caspi and his coauthors hypothesized that the genetic variant would moderate the effects of stressful life events on later depressive symptoms. Using the same Dunedin data set that was deployed in the *MAO-A* study, the researchers found evidence consistent with the animal models: those persons whose genetics suggest they will be sensitive to stressful environments and whose environments had been stressful evinced more depressive symptoms than those with similar environments but who possessed the stress-free genetic variant.

Almost single-handedly, this research launched a direction of inquiry mapping the interplay of biological and environmental predictors onto important outcomes. These two papers have received nearly ten thousand scientific citations in the last decade or so and have been the subject of substantial interpretation, counterinterpretation, statistical prodding, replication attempts (with many successes and as many failures), and summary analyses in the literature. The debate continues, with competing summary analyses (called "meta" analyses) reaching different conclusions.[19]

THEORIES OF GENE-ENVIRONMENT INTERACTION

We need a theory of gene-environment interplay in which to situate the exciting findings of Caspi and his colleagues. Central to a good theory of the salience of gene-environment interactions is addressing several questions. Why would human genes "want" to be contingent upon the environment? Why is this sensitivity to context useful for the survival and thriving of the species? Isn't it risky to allow the (uncertain) environment to dictate human outcomes? Or is it more risky to have genes functioning completely independently from the environment?

An early framework to conceptualize gene-environment interaction was the diathesis-stress hypothesis, which suggested that some individuals are born with "risky" genetic variants and others are born with "safe" genetic variants, but both individuals would have about the same outcomes if placed in a neutral environment. If, however, individuals with "risky" genetic variants were placed in extreme environments, we would see disproportionately different outcomes—a gene-environment interaction. Indeed, there is quite a bit of (still-controversial) evidence that children who are abused and have "risky" gene variants related to the serotonin system in their brains experience incredible reductions in life outcomes, whereas abused children with alternate gene variants are worse off, but not multiplicatively so.[20] Why would humans have "risky" genetic variants at all, if there were no benefits? Why would evolutionary pressures fail to wipe out these variants? A couple of theories have been proposed. The first set of theories is that these "risky" variants were, in the not-too-distant past, beneficial. Perhaps the genetic variants that we see producing "overreactions" to current stressful environments would have been excellent variants to have in the Rift Valley, but humans are too recently displaced from these environments for the variants to have disappeared through selection. The "warrior" gene might have been good for populations who need warriors to fend off lions to survive. Or perhaps these "risky" variants are risky for some outcomes but actually protective for other, often unmeasured, outcomes. Schizophrenia is often positively associated with measures of creativity and intelligence, for example.[21]

A second theory considers the gene at the level of the species rather than at that of the individual and supposes that there are variants that are "orchids" and other variants that are "dandelions." The dandelions manage in most environmental circumstances. The orchids thrive in ideal environments but wilt in others. The dandelions are the group's way of hedging its bets against changes in the environment (though group selection remains a very controversial notion in evolution). At the level of a species, humans would like to have both types of genes in their pool, as a massive insurance policy. In case of large (or catastrophic) changes in the environment (e.g., climate change) that might wipe out all the orchids who are fine-tuned to the current environment, having dandelions in the population could be essential for the survival of the species.

This sort of evolutionary risk management has been found in a wide variety of species, from *Arabidopsis thaliana* (mustard weed), with regard to germination time,[22] to *Saccharomyces cerevisiae* (yeast), with regard to growth and replication strategies.[23] In the case of *Arabidopsis*, one (genetically driven) "decision" is the timing of the sprouting of a seed in order to make the best use of environmental conditions (rain, sunlight, and so on). If the idea is to sprout on the day with the most sun, for example, then all plants in the northern hemisphere should sprout on the vernal equinox (approximately June 21). But what if that day happens to be extremely cloudy or have a solar eclipse that year? It might behoove a population of *Arabidopsis* to have some individuals not so carefully regulate their germination timing—in other words, the group could trade maximal payoff for a range of germination times as a form of insurance. In this case, the group that bets it all on June 21 are the orchids and those that add a bit of randomness to their germination date are the dandelions—they are not as affected by the cloud conditions on one particular day. In stable, cloudless conditions, the orchids should win out since they will dominate the landscape by timing their germination perfectly. But in a more unstable situation (think global climate change), the species with dandelions to hedge their bets may predominate. Natural selection will find the right balance to match the variability in the environment.

YET ANOTHER CHALLENGE . . .

As if it were not complicated enough to examine these ideas with the data we have available, a group of social scientists has pointed out potentially fatal flaws in some of the gene-environment interaction literature.

In order to properly explore gene-environment interactions, we have another issue to think about: whether the environment in which we are interested is under human control (subject to the research subject's choices) or whether it is "exogenous" (unrelated to an individual's actions and choices, and, especially, genotype). This issue is important because the entire enterprise for assessing the interplay between genetics and environments is predicated on two key steps: first, separating the components of the social processes that are "genetic" from those that are "environmental," and second, exploring how they are mutually dependent. The first step is more difficult than it sounds. Rather than a clear distinction between actions and factors that are genetic or environmental, what if the environment we experience is partially driven by what gene variants we carry around? Our genes might shape which environments we interact with, rather than genes and environments being two separate forces that act on us, through both their immediate effects upon us and their interactions with each other. The situation in which genotypes affect the environments we experience (genes select environments) is called gene-environment correlation and comes in several versions.[24]

A first source of gene-environment correlation is *active* gene-environment correlation. People (children and adults alike) have some agency over the type of situations they place themselves in—not all situations of course, but some—and they may do so in response to how they "fit in" with or gain from their environment. This process is called selective gene-environment correlation. For example, people who are extroverted may seek out different environments (friends, parties, and so on) than people who are introverted. Alternatively, rather than selecting an environment, genes may elicit an environmental response—labeled evocative-gene environment correlation. For example, children who are disruptive (in part based on genes) may generate (evoke) punitive reactions (environments) from parents and teachers and other adults and children.

To give another example, if children with the A variant at a locus in a gene we will call *SCHOOL* are more likely to gain entrance to high-performing schools than children with the T variant, children with the A variant will attend better schools than children with the T variant. If the A allele acted only by affecting school choice, this would not be a problem. It would suggest a world in which the mechanism by which the A allele had its effect on a downstream outcome—say wages—worked through school choice. But now imagine that the A allele for *SCHOOL* has two effects. First, it affects school choice. Second, it affects wages by raising IQ. This dynamic might fool a social scientist who analyzed the effects of schools to falsely conclude that "good" schools were causing certain outcomes when it was really the pleiotropic (multipronged) effect of the *SCHOOL* gene affecting *both* school and wages. That is, in the counterfactual world in which we made the Ts go to the so-called great schools, they might not experience any wage gains because the effect of schools was really a statistical artifact of the genetic sorting going on.

Our ability to infer cause and effect becomes even more complicated when multiple genes affect an outcome. Imagine, as in the case above, As on the *SCHOOL* locus select into better schools than the Ts do, but there is no direct effect of the *SCHOOL* gene on IQ or wages in this new scenario. But then add a second gene locus that has two variants: C and G at the locus called *COGNI*. Let us posit that there is a gene-gene interaction effect such that for Ts at the first, *SCHOOL* locus, variation in this second, *COGNI*, gene has no effect. Ts at the *SCHOOL* locus get bad educations, test low on cognitive assessments, and earn lower wages no matter whether they have C or G at *COGNI*. But imagine that there is a specific combination that yields huge effects. As at the *SCHOOL* locus who have Cs at the *COGNI* locus have higher IQs and wages than As who have Gs at the *COGNI* locus. That is, we see an effect on wages of *SCHOOL* only in the presence of the C allele in *COGNI*. We can turn this around and say we only see an effect on wages and IQ of the *COGNI* gene variation when we have A at the *SCHOOL* gene. If we did not measure the *SCHOOL* gene explicitly and look for this genetic interaction, a scientist studying only *COGNI* might falsely infer that there was a gene-environment interaction in which the C allele of *COGNI* only raised IQ and wages when a child with that allele went to a good school. In actuality, because of gene-environment correlation, the measurement of school quality was actually, secretly, acting as a genetic test for the A allele on *SCHOOL*! That is, an unmeasured gene-gene interaction was masquerading as a gene-environment interaction.

When we think intergenerationally, things become more complicated still. Since children are genetically related to their parents, and parents select their children's neighborhoods, schools, and to some extent their friends, sports, movies, and stressors (at least indirectly),[25] we have a problem in analyzing gene-environment interactions. When parents make these important choices about what their children's environments look like (perhaps based on their own *SCHOOL* alleles), it is conceptually (and statistically) as if the children's genes are helping to select their environments, although doing so passively—a so-called *passive* gene-environment correlation. Since the parents pass both the *SCHOOL* and *COGNI* genes on to their children, even when children have no apparent individual choice about the environment in question (since the parents decide what neighborhood they live in, how much income they have, and so on), an apparent gene-environment interaction may really be masking a gene-gene interaction because parents pass on both the environment and their genes.

So, the exposure to maltreatment in Avshalom Caspi et al.'s original studies[26]—the "environmental" measure—may in fact be capturing a component of the parental (and by extension, child's) genotype, muddying our ability to separate the two and analyze their interactions. This is a particular worry for Caspi et al.'s analysis because it was most likely the parents who were abusing or neglecting the children. Was it the environment that certain parents were creating (abuse) that interacted with the *5-HTT* variant to produce depression in the child? Or was it that another gene—let us call it *ABUS*—caused the parents to act that way? If the latter, we can hypothesize that when the risk variant of *ABUS* is inherited with the short version of *5-HTT* promoter, those two genes interact to produce depression in the child. But when a child receives only the short allele(s) for *5-HTT* (and not the *ABUS* gene), then we see no effect on her probability of experiencing depression. *ABUS* and *5-HTT* would together form part of a gene-gene interaction. If we do not measure *ABUS* explicitly (or even know where to look for a potential partner gene to *5-HTT*), then we might mistakenly assume the parentally created environment was what was interacting with *5-HTT* in a contingent way, and not the unmeasured gene that was driving the creation of the parentally evoked environment.

Parental genotype (and child genotype) may therefore be predictive of the risk of perpetrating (and exposure to) child maltreatment in that the genes are selecting (predicting) environments rather than merely interacting with them. Thus, we cannot conclusively distinguish whether the Caspi et al. study is capturing a gene-gene interaction or a gene-environment interaction. In the first case, the child's measured genotype would be interacting with the child's unmeasured genotype (i.e., other alleles that may influence behavior but are not measured explicitly in such a candidate gene approach). This can occur because the unmeasured child's genotype (e.g., the *ABUS* allele) is aligned with the parent's genotype, which is then aligned with the parent's risk of mistreatment of her children. In the second case, we are capturing a gene-environment interaction in which the interaction is only between the child's genotype and her environment.

An important implication of the existence of gene-environment correlations (regardless of the type) is that their existence challenges our ability to uncover gene-environment interactions. We cannot be sure whether we are uncovering gene-environment interactions or gene-gene interactions. If genes affect environments through gene-environment correlations, then it could be that we are estimating the interaction of the "behavior genes" with the "environment-selection genes" and not uncovering anything particularly informative about the environment per se. This issue has proven to be an important problem in the gene-environment interaction literature and may be part of the explanation for why some of the earliest and most notable studies that provided evidence for gene-environment interactions have failed to be replicated and are now disputed.

The issue of separating genetic from environmental factors has a long history that predates the modern (molecular) analysis of gene-environment interactions. For example, in 1980, Christopher Jencks focused on interactions in IQ investigations and called attention to mental retardation.[27] It had been well established that phenylketonuria (PKU) is a human genetic condition caused by mutations to a gene coding for a specific liver enzyme. The absence of a functional enzyme disturbs brain development, leading to mental retardation. But with modern screening programs and a simple dietary adjustment, babies with PKU can be spared this cause of mental retardation. An environmental intervention can transform the probability of inheriting a genetically caused predictor of low IQ from very high (with no environmental/dietary intervention) to essentially zero. So, does that mean that IQ is genetically or environmentally determined in this pathway? For those without PKU, there is no effect of environment (the special diet) on IQ. For those with the dangerous genetic variant, the environment has major effects. In general, how can we ever hope to separate genetics-only factors from environment-only factors?[28]

There is a small but growing literature that has attempted to separate gene-environment correlation and interaction by conducting studies that focus attention only on environmental variation that is not likely the result of (or correlated with) genetic variation.[29] The day on which you were born probably has a very limited relationship to your genotype but also probably matters not a whit for your life outcomes. Or does it? Sometimes the day you are born on shapes your whole life.

As those of a certain generation will always remember, during the Vietnam War there was a lottery draft in which men of certain birth years (i.e., 1950) were given a draft number based on their specific birthday (e.g., July 22). So now we have a birth day that means something but is still unrelated to genotype.[30] We used this information to ask whether being assigned a number that increased the risk of serving during the Vietnam War interacts with genotype variation related to smoking to predict smoking status during adulthood. Does military service—with its stresses and high degree of exposure to tobacco—differentially elicit responses based on genetic predisposition to smoke?[31] We found evidence that veterans with a high genetic predisposition for smoking were more likely to become regular smokers, smoke heavily, and be at a higher risk of being diagnosed with cancer or hypertension than nonveterans. Military stress seemed to interact with genetic risk for tobacco use. And this evidence is "clean"—it is not affected by gene-environment correlation because genetics does not affect whether you drew a high or low number in the Vietnam lottery.

Another example of the new focus on "clean" evidence of gene-environment interactions, coincidentally, also uses an important date. That date is memorable to not just a given generation, but to pretty much anyone who is alive today: September 11, 2001. Like our use of birth dates in the Vietnam draft lottery, we used 9/11 as an important date in our effort to find environmental exposures that are both important and also unrelated to genotype.[32] As we were beginning the research, we realized that a large national study was being conducted in August 2001, September 2001, and later—a study that obtained subjects' saliva for DNA analysis and asked the donors whether they were experiencing depressive symptoms. The study had been tracking thousands of people for more than five years, beginning when they were in high school. Our research approach used this coincidental timing of the survey to compare people who had completed their survey before or after the terrorist attack on 9/11 (e.g., comparing interview responses in August 2001 with those who were interviewed post-9/11) to assess whether genotype (the long alleles of the *5-HTT* promoter, in particular) was associated with differential responses to the attacks with respect to depressive symptoms.[33] This research is related to the second Caspi et al. study in *Science* that examined interactions between stressful life events and *5-HTT* promoter length variation in predicting depressive symptoms.[34] However, we found results inconsistent with this older study. While Caspi et al. found that the short/short genetic variant increased the effects of stressful life events on adult depressive symptoms, we found the opposite pattern: the short/short variant reduced the effects of 9/11 on depressive symptoms. The key is that who was interviewed before or after 9/11 was random.

One possible reason for the different pattern of findings from those of Caspi et al. is that we designed a study that was able to separate gene-environment correlation from gene-environment interaction processes.[35] What are the assumptions needed to infer cause and effect in the two studies? Each study requires that genotype be unrelated to the exposure (i.e., no gene-environment correlation). For our study, this condition specifically means that genotype must be unrelated to whether someone happens to be interviewed before 9/11 or after 9/11. For the Caspi et al. study, it means that genotype must be unrelated to whether a child is the victim of maltreatment. In our view, the assumption needed for our research design is more plausible.

Studies that plausibly separate genetic from environmental effects are crucial to understand gene-environmental effects but are rare in the literature. This is partly due to the catch-22 involved. Researchers' attempts to unearth environmental factors that are both important explanations of socioeconomic phenomena and are also unrelated to genetic factors have been extremely frustrated by the difficulty of finding cases that fit both criteria.

Critics of this "natural experiment" approach dismiss these "pure" studies as novelties that fail to describe the world, noting that we cannot rely on draft lotteries and terrorist attacks to understand the most compelling questions in gene-environment interaction research. This is a version of what some people call the problem of extremely "local" treatment effects. Let's take an example from the education literature. Many economists have used the expansion of compulsory schooling laws in U.S. states as a natural experiment to find out what the "true" causal effect of an additional year of schooling is on wages or life expectancy.[36] In this research design, the additional year of formal education is distinct from the background factors that may themselves cause students to acquire more education but that may also directly affect earnings or mortality (say IQ, perseverance, or any number of factors). However, since the impact of the laws was only to increase the marginal child's formal education from 10 years to 11, what we learn from the examination is limited. We learn the effect of that particular transition (from 10 to 11 years of schooling) on children who otherwise would have left school after 10 years. Examining compulsory schooling laws does not tell us about the effect of that additional year (junior year) on the type of student who would have stayed regardless of the law. These laws also do not tell us the value of an additional year of schooling at another point on the educational ladder. If we are interested in the effects of college attendance on adult earnings, analyzing the change in required formal education from 10 years to 11 does not necessarily help us. The effects we uncover for the transition from 10 to 11 years of schooling may indeed tell us something about other educational transitions, if we think the effect of education is the same for all children and if we think it is the same for each additional year—but, obviously, these are big assumptions.

While considering the potential problems of extrapolating local treatment effects to broader questions of interest, we also need to consider the problems with other approaches. One of the principal alternatives to focusing on "local" effects that we believe are accurate is to focus on estimates that we think are biased. That is, if we are interested in the effect of the transition from 11 years of schooling to 12 years of schooling, we face a choice. We can either use the knowledge gained from "clean" evidence generated by compulsory schooling laws and the resulting transition from 10 to 11 years of schooling, or we can use results from research designs that do not produce "clean" evidence but do directly examine the transition from 11 years of schooling to 12 years. Of course, we could try to combine both kinds of evidence in some way, but what if they yield different results?

In addition to their narrow focus (i.e., estimating the effects of 11 versus 10 years of schooling rather than the effects of all years of schooling), natural experiments also fail another test in science—replication. An important scientific standard of evidence often used in genetic analysis is whether researchers are able to produce the same finding across a variety of datasets, times, and places.[37] Draft lotteries and terrorist attacks on the scale used in these papers are, thankfully, rare, so do not allow the replication efforts that are required in some scientific disciplines. The difficulty of replication also raises the question of whether the findings generated from the study have external validity or are highly local: whether the same results would be expected if the draft lottery occurred in 1949 or 2009 rather than 1969—or any other time external to the study period. Another issue with many of these "pure" studies of gene-environment interactions is the reliance on candidate gene approaches to measure G. As we described in chapter 3, many candidate gene findings have been false positives. Using the candidate gene approach for gene-environment research further increases the probability of false-positive findings (because not only are the number of potential outcomes in the thousands, the number of environments to test is also very large). One initial response to the issues with candidate genes has been to use polygenic scores instead,[38] though the concern that polygenic scores are a black box remains.[39] Ideally, the way forward is to merge the advantages of each approach while acknowledging the limitations of each; there is no single perfect analysis.

LEVERAGING GENE-ENVIRONMENT INTERACTIONS FOR BETTER LIVES

We have been promised for decades that the medical establishment in general, and pharmaceutical manufacturers in particular, are on the cusp of tailoring medical services and medical treatments to fit our individual needs. Medical genetics was meant to be the way of doing just this; medical practitioners could, with a sample of saliva, determine why a patient was obese or experiencing symptoms of depression. This technological advance is important, because nearly all weight-loss diets fail, and prescriptions for diagnoses like depression are trial-and-error enterprises. Clinicians prescribe medications based on many factors: experience (with other patients), what little information they can gather about the patient in a seven-minute office interaction, other intuitions, and drug company enthusiasm for a product (through detailing and free samples).[40] And clinicians often guess wrong. They prescribe the wrong drug, they may miss interactive side effects with current drugs, and, most basically, they are not able to align drug choices with the genetic disposition of the patient. The result is that after waiting for weeks for the (wrong) drug to take effect, patients are forced into another office visit followed by another step down the trial-and-error decision path.

A new direction of so-called precision medicine seeks to overcome the past reliance on trial-and-error prescribing, in particular for depression, alcohol abuse, tobacco use, and obesity, among other health behaviors and conditions. The underlying determinants of these conditions vary greatly across a population. Many people experience depressive symptoms following a tragic event, others have seasonal affective disorder (SAD) and may experience sensitivity to extended periods without sunshine (think Seattle),[41] and still others could be responding to pubertal changes during adolescence. And below the surface of these sources of symptoms lie genetic vulnerabilities.

Not everyone who experiences a tragic event suffers from depression, and not everyone in Seattle suffers from SAD. It is reasonable, then, to consider whether the variety of genetic and environmental sources of depressive symptoms might be treated with more targeted medications and therapy. Many drugs for depression target different biological pathways (and we often do not know the precise mechanisms). Effexor may work by blocking serotonin and norepinephrine reuptake by the presynaptic neuron;[42] Wellbutrin may work by targeting nicotinic acetylcholine receptors;[43] and Lexapro may work by more specifically targeting serotonin.[44] They each have different side effects, too (suicidal thoughts for [young] Effexor users, seizures for Wellbutrin users, and decreased libido for some Lexapro users, among other side effects). Might it make sense to measure well-studied polymorphisms in the dopamine (*DRD2, DRD4*, and *DAT*) and serotonin (*5-HTT*) systems and to divide patients into groups before deciding which drug might work best for them, rather than basing the decision of which to start with on the doctor's own proclivities? Efforts are underway to better match the likely biological and environmental pathways with treatment choices.[45]

Similarly, the reasons that people start (and continue) to smoke vary considerably. Nearly everyone in the United States who has ever really "started" to smoke (often defined as smoking 100+ cigarettes in your life) has done so between the ages of 14 and 22. Most start during high school, but some who never smoked during high school start during college or on their first job. But almost no one who is 30, 40, or 50 has started smoking recently. In contrast to the very clear age patterns of smoking initiation, the patterns of quitting are quite different. For many people, successful quitting is a lifelong effort, with many failed attempts. Most smokers have tried to quit, often many times.

This age profile of smoking is at odds with many other health behaviors and might be one reason for the difficulty of successful treatment. That is, for most people, starting to smoke is an intensely social phenomenon. People smoke to conform to their friends' behavior and popular culture. And most (in the United States) probably do not initiate smoking primarily due to genetic propensities to smoke. After all, even if they are genetically wired to crave nicotine, they cannot know this until they take their first drag.

While most people experience smoking initiation in a similar way, smoking cessation is very different; some smokers appear to be able to quit cold turkey, and others need multiple rounds of treatment regimes. Unlike initiation, the determinants of smoking cessation are probably reversed, social factors playing a subservient role to genetic factors. No doubt, having friends, coworkers, or a spouse who smokes will make it more difficult to quit, but there is much evidence detailing the genetics underlying successful quitting. This is not to say that the ability to quit is "determined" by genetics and that some people should just give up trying. But, like treatments for depression, being able to match the underlying biology for unsuccessful attempts to quit with a treatment that targets these factors may enhance success. Stated differently, harnessing gene-environment interactions—choosing the "environment" (i.e., treatment) strategically—could be a new path toward reducing smoking rates.

New approaches to matching patients to depression medication and tobacco cessation strategies are just two examples that compose the burgeoning field of precision medicine, in which clinical-trial data is being mined for hints of potential gene-treatment interactions, and new clinical trials are being undertaken with the explicit purpose of targeting genetic moderators of treatment to understand why treatments succeed for some people but fail for others.

In one so-called pharmacogenetic clinical trial[46]—a clinical trial that specifically focuses on how the effects of treatments may vary according to a person's genetic code—investigators have examined the effects of receiving a nicotine patch or a nasal spray or bupropion (an antidepressant also used to treat nicotine addiction) compared with a placebo group. In examining the ability to quit during the following week after treatment, the researchers found that individuals with a specific genotype (at least one minor C allele in a *GALR1* polymorphism, rs2717162) who were given the bupropion drug had about a 50 percent lower chance of quitting. A genetic variant associated with nicotine cravings was able to render a treatment ineffective. However, individuals with this same genotype using the nicotine patch or nasal spray were equally able to quit as people with a different genotype using the same methods. These results suggest that individuals with that particular genotype should not take bupropion as a first-line treatment.[47]

Extending findings from gene-environment-interaction studies into clinical practice may lead to more successful attempts at quitting. But without letting genetic information help guide some aspects of medication practices, many individuals will continue to fail at quitting, perhaps unnecessarily. We use tobacco use as a key example here in part because of its public health consequences around the world. In fact, the World Health Organization places tobacco use as the single greatest cause of preventable death in the world.[48] But new research will no doubt extend these examples to a much broader range of outcomes—from obesity and diet to alcohol abuse, drug addiction, and beyond.[49] Indeed, the usefulness of targeting treatments based on genotype becomes more apparent as we think about the different treatments available for smoking cessation, depression, and other health issues.

Although taking advantage of genetic differences when targeting medications and nutrition to individuals is controversial, the controversy is anchored in the underlying science—specifically, whether the current state of the science is conclusive enough to be useful—more than in the underlying ethics. An interesting and important set of follow-up questions will involve extending the promise of personalized *medicine*, with its focus on the molecular targets of drugs, to personalized *policies*, which could include the much larger set of interactions that people have with governments and agencies, from tax policy to education funding.

A case in point is the example of tobacco taxes in the United States. The Institute of Medicine and Centers for Disease Control and Prevention, among others, have ranked this policy as one of the top ten most impactful for increasing public health in the past century. This is because the United States has witnessed enormous changes in smoking since the first taxes were introduced following the surgeon general's report in the mid-1960s. Smoking rates have been cut in half. And raising taxes is cheap for the government, so the benefit-cost ratio is incredibly favorable.

But the last decade has seen a reversal. In the face of the largest to-bacco tax increases in our history, tobacco use has remained virtually unchanged. Have the laws of economics been repealed? New evidence that combines genetics and social policy evaluation has pointed to an explanation: the pool of smokers in the 1960s was genetically dif-ferent than the pool of smokers now. People carry different variants of a group of nicotinic receptor genes, and as the name suggests, these genes influence how much dopamine (a pleasure chemical) is released when people smoke; they affect how much one "likes" smok-ing. These genetic variants also interact with smokers' environments. One hypothesis is that many of the smokers in the 1960s did not need a big push (tax increase or social pressure) to quit smoking (or to never start in the first place). But over time, these small pushes have mostly caused people to stop smoking if their genes did not put up a big fight. And what we have left in the smoking pool is a dispro-portionate number of persons whose genes are fighting back, partly because nicotine is so pleasurable. Our evaluations of tobacco tax policy suggest that only adults with a low genetic risk of smoking will still respond to taxation, and adults at higher genetic risk will remain unmoved.[50]

What does this mean for policy? Should we keep increasing taxes—forcing the remaining smokers to pay a larger and larger share of their incomes in part because they were unlucky in the genetic lottery of life? Should we subsidize their smoking because they love it so much (due in part to their genetics)? Or should we target these folks for prevention efforts so they do not start in the first place?

As with smoking in adults, another problem that at times seems in-tractable but also has seen a lot of important progress is the birth out-comes of children. One of the most heavily documented and tracked measures for babies is birth weight. It is easily measured, squirm-ing aside, but is also considered a good summary of how healthy the infant is in its first moments of life. It has considerable long-term impacts on life outcomes, and the fact that the United States lags behind much of the developed world in reducing our rates of low birth weight suggests the need for further investigation.

Even among identical twins, whose genetics are the same, the heavier twin has been shown to excel in life by a variety of measures. Some of the most consistent evidence concerns health outcomes and cognition, especially IQ. Sandra Black and her colleagues used Norwegian registry data on a very large sample of identical twins to show that a 10 percent increase in birth weight (holding everything else constant) is associated with a 5 percent increase in IQ around age 18.[51] Low-birth-weight siblings are significantly less likely to graduate from high school in a timely fashion than their normal-birth-weight siblings according to our own earlier work.[52] Other researchers have shown that birth-weight differences affect outcomes even later in life, including the development of type II diabetes and markers of cellular aging (i.e., telomere length).[53]

However, like the findings from Caspi et al. on maltreatment in children, findings on low-birth-weight babies suggest that resilience is also an important part of the story—not all low-birth-weight babies grow up to have lower IQs and worse health. Indeed, our research shows that genetic measures of neuroplasticity (the ability of the brain to change in response to the environment) seem to shape children's ability to overcome low-birth-weight status with respect to future outcomes. Using a summary measure of three genes (*BDNF*, *COMT*, and *APOE*) available in a representative survey that has followed more than 10,000 Wisconsinites (who graduated from high school in 1957) for over 50 years, we find evidence that higher birth weights are related to higher IQs for some individuals but not for others.[54]

The orchid-dandelion distinction appears to be at play here. Higher birth weight is good for most people (up to a certain point, of course), but the (genetic) dandelions barely respond at all to birth weight in determining IQ as an adult. However, genetic orchids are a different story—they wilt when facing lower birth weights and thrive when enjoying higher birth weights. And these effects stay with them. Not only are IQs affected during late adolescence among this group of genetically sensitive individuals, but so also are adult wages and employment outcomes, which were measured when the sample respondents were more than 50 years old.

Like the emerging evidence from differential responses to tobacco taxes (based in part on genetic factors), there are related policy possibilities for birth weight. Conventionally, hospitals would be on the front lines of this issue, because they measure birth weight and also genotype (or could). They typically collect blood from new babies through heel pinpricks. It would only take one extra step to assess infants for genetic measures of neuroplasticity, and to then implement a more targeted approach to selecting who receives which type of interventions so that we reduce the impact of low birth weight on longer-term outcomes. Nowadays, infants who are born with a low birth weight are provided a range of interventions (e.g., nutrition supplements, hospital monitoring, additional visits to the pediatrician), many of which have been judged to be relatively ineffective in terms of reducing mortality rates.[55] But the orchid-dandelion results above suggest two important implications. First, since many babies who receive the interventions for having low birth weights are dandelions, it becomes clearer why the impacts of interventions are small or nonexistent. Second, if we could somehow target the interventions toward babies most likely to benefit from them, the impact per dollar spent (cost-effectiveness) would substantially increase.

This new style of "personalized interventions" is (and will remain) highly controversial, but it is also likely to become more useful in the not-so-distant future. Like the issues with targeted (personalized) tobacco tax polices above, there are ethically murky waters in any sort of genetic targeting. Will we actually withhold interventions for low-birth-weight babies who genetic models predict will be unaffected by the interventions? How accurate a prediction will be necessary for us to consider withholding care? What if the predictions vary by race or ethnicity (which is very likely)?

Another question for the "personalized" world is scale. Some interventions are more difficult to personalize because they are governed by state or federal laws and policies but at the same time are very personal issues in the lives of millions of people. Social programs, especially income transfers such as welfare, have a mixed record of impact on the success of poor children, although many have made modest gains in increasing children's achievement and life outcomes. On the one hand, we have had small-scale successes, such as the Perry Preschool Program, a program in which a highly targeted group of disadvantaged African American children from Michigan in the 1960s received preschool education and care services. The program was expensive and highly intensive but had lasting effects on a host of outcomes for the children, ranging from reductions in criminal activities to better health outcomes forty years later. On the other hand, broad-based (i.e., less targeted) programs like the earned income tax credit or cash welfare have had mixed (but typically much smaller) effects on children as they grow up.

Within the literature on gene-environment interaction, researchers are beginning to ask whether income—and reductions in material hardship more generally—affect children differentially based on their biological and genetic "endowments." Indeed, preliminary evidence is emerging to support this view. A young economist, Owen Thompson, has recently published evidence from a national sample of adolescents in the United States suggesting that the "warrior gene" may be partially responsible for reducing the impact of family income on future schooling outcomes.[56] Males in the sample completed more schooling, on average, if they were from a richer family. Thus, an environmental exposure (family income) that can be changed through policy predicts children's educational attainment. Although this relationship between family resources and children's educational success is well documented,[57] Thompson asked a new question: are the effects of family income different if the child has one genotype or another? Indeed, his findings suggest that while more income was better for the educational success of all boys, for boys with the low-activity (risky) allele of MAO-A, family income was particularly important. These boys gained approximately three times as much benefit from being in a richer family than those with the high-activity allele. Of course, this study suffers from the same cloud of endogenous environments and potential gene-gene interactions obfuscating the causal scenario as the original Caspi et al. study did, but it is a very useful starting point in understanding potential differential effects of income on children's outcomes.

Digging deeper into the mechanisms linking genotype, family resources, and child outcomes, other researchers have focused attention on whether family stress is an important pathway. A group of sociologists, economists, child developmental scientists, and geneticists have linked the impacts of the Great Recession to measures of "harsh" parenting.[58] Economic stress makes all parents more likely to be irritable and short-tempered (thus more likely to snap at or hit their children). But in this study, the authors found that a particular gene variant in the dopamine system (*DRD2*) that some mothers carried made them *much* more likely to yell at or hit their children during periods of economic stress. These *DRD2* at-risk moms behaved similarly toward their children as other mothers when economic conditions were good, but when the economy tanked, their risky allele manifested an effect. These mothers reacted more strongly to poor economic conditions than other mothers did, and they may pass the negative effects of economic hardship—through their harsh parenting practices that are elicited by the stressful environment—to the next generation.

Expansive and expensive economic policies aimed at supporting the incomes of poor families are currently targeted based on income needs ("means tested") rather than targeted based on the likely impacts on the parents and children. This emerging evidence, like the evidence for interventions on birth weight, suggests the possibility of further targeting on the basis, in part, of genetic factors. That is, if we can use genetic information (e.g., genes related to neuroplasticity) to construct statistical predictions of which low-birth-weight babies may benefit from additional hospital care and which may not, we could also use genetic information to predict which children may benefit more from family income support programs and which children may not. Using information about the "warrior gene" examined by Thompson is an example, but using genome-wide data could create better predictions. These ideas may seem "out-there" and are unlikely to be put into practice any time soon, especially because most of the studies we reviewed are based on candidate genes and have limitations in the research designs that make the results suggestive but not conclusive. The data limitations, however, are quickly disappearing, so we need to shift attention from whether we *can* create predictions to whether or not we *should*.

The reality of genetically informed social policy may be very discomfiting because it could mean that, in the interest of maximizing efficiency, we withhold treatment from some people based on the output from predictive statistical models of their probable (small) effects. The National Institute for Health and Care Excellence (NICE) in Britain already creates guidelines for appropriate clinical care based on cost-effectiveness studies. So if the health benefits of a procedure are too costly (the British are willing to spend about $30,000 for an additional quality-adjusted life year), then the procedure is not recommended (and not financed by the government).[59] But NICE looks only at the average benefits of a procedure. What if a procedure produced enormous benefits for a small number of people and nothing for most people? And what if genetic analysis could predict whether a specific person was in the small group who would benefit or the large group who would not?

While it may be disturbing to think that we would give or withhold treatment or resources (or policies) based on someone's genetic code, we might consider that we already ration health care, access to good schools, and many other resources based on the ability to pay. In fact, since educational attainment, employment, and many markers of the ability to pay have some genetic roots, our current system of price-rationing access to many goods does have a hidden component of rationing based on genetics.

Instead of using only the ability to pay, an alternative allocation rule might be based on the ability to benefit from the resource. There is no doubt that the effects of a vast array of environmental factors, including medical treatments as well as social policies, have different effects on different people. Some of the sources of these different effects will no doubt be associated with easily measurable demographic or environmental factors like age, gender, or income level, but others may depend on genetic variation. Currently, we make no real attempt to leverage such genotypic variation to maximize the benefits from each treatment and policy (while we do allocate certain resources, like Medicare and Medicaid, based on age and poverty, respectively). Instead, we mostly rely on a belief (from economists, mainly) that those who will benefit the most from a resource are those who will pay the most to purchase the resource. This can work in some cases, particularly when poor households can borrow easily to finance investments that they would especially benefit from; however, without equal access to financing, information, and social connections, poor households can be closed out from leveraging resources. This economically based rationing approach also assumes individuals can accurately assess the potential returns on investments. The ability to use genetic information to predict responsiveness to different treatments and policies is in our immediate future once the relevant polygenic scores are improved; it is imperative for us to consider whether and how to use this information.

CONCLUSION: WHITHER GENOTOCRACY?

One effect of the social genomics revolution that is not yet fully felt is how people will deal with and understand genetic information about themselves and their loved ones. Now that genetic data are becoming widely available, how will this new knowledge be used by the larger populace—that is, big data meets big public? Who will translate this so-called language of life for nonspecialists—those of us who know only snippets of the relevant words and concepts and who have not thought about biology or genetics since high school or college? How do we sift through the clutter of raw data, false positives, and real risks, especially when much of it is in a language that most of us do not understand?

Nowadays, you can spit into a cup, mail the contents to 23andme or a similar company, and enter your credit card information online. About four weeks and a hundred dollars later, you will have access to one million bits of information about your genome, such as whether you have the genetic variant TA or AA or TT (a million times across your genome). For nearly everybody, this unfiltered information is completely useless. It is similar to trying to read the ones and zeros of a digital computer file for the first time. The ones and zeros tell you what a machine (your cells and body) is programmed to do, but no one knows yet how to read the complete text to understand what the program will do under different circumstances. Geneticists have assigned phenotypic relevance to some of these data, but otherwise they remain a mystery. In the early days of the industry, 23andme and similar companies would also send you an assessment—translating these TAs, AAs, and TTs for you. They might send you pie charts, line graphs, and summaries such as "your odds of having a stroke are 20 percent higher than average"; "your odds of developing Alzheimer's disease are 18 percent higher than average"[1]—scary stuff.[2] But the FDA has ruled that these assessments are basically junk.[3] For a number of years these companies could only send you assessments about your genetic heritage—that you may have distant cousins in Kenya (or Texas) or have a lot of Neanderthal DNA in your genome. (However, in February 2015, the FDA decision was partially rolled back by allowing companies to describe specific genetic risk factors on a smaller class of inherited health conditions. Stay tuned.[4])

But this regulatory hurdle is really just a hiccup. It will not be long before you can hire companies to take your genetic data and create polygenic scores for anything—from educational attainment to BMI to entrepreneurship proclivity to risk of depression. And these scores will become more accurate over the next decade in their capacities to predict. Right now a polygenic score predicts about 6 percent of the variance in educational attainment among populations of European descent, but some think it will get close to 20 percent (or higher) over time. (Even with low predictive power, the effects are large—analogous to the way the dangerous version of the *BRCA* gene explains little of the variance in the population prevalence of breast cancer but still predicts an eightfold difference in lifetime risk for a woman of contracting the disease.) Still, we currently have almost no understanding of how the genetic measures underneath these scores work to produce outcomes such as years of completed schooling or BMI. Given their grab-bag nature, these scores sacrifice a detailed understanding of biological (or social) processes. Compared with the much maligned candidate gene assessments, polygenic scores instead maximize raw predictive power.[5] So, this information is potentially not very useful for your life *now*. It is as if we tell you that being born in June is associated with lower schooling attainment[6]—it's too late to change your birth month. (But perhaps not too late to plan your as-of-yet unconceived child's birth date.)

THINK OF THE CHILDREN

Although the information we are gathering might have little value now, what about your kids and your *potential* kids? Currently, clinical prenatal genetic diagnosis typically amplifies only enough DNA to conduct a chromosomal scan or to probe for specific mutations that children may be at risk of inheriting. But this technological problem is being solved in research laboratories, and soon clinicians will be able to extract a few cells from five-day blastula and amplify enough DNA to read the entire genome. What happens as we move from a twelve-week prenatal blood test that searches for "clear" markers for Down syndrome or other chromosomal anomalies and an eighteen-week ultrasound that can determine the sex of the fetus to a genome-wide test that reports eye color, predicted height, predicted BMI, predicted IQ, and even predicted income of the embryo shortly after conception (perhaps even before implantation, in the case of in vitro fertilization)? What happens if this test is not covered by insurance? Limited versions of this are offered in Australia for $1,000.[7] Blood tests are used to scan the mother's blood for fragments of the fetus' DNA that can, in principle, be sequenced, but the test is presently aimed only at providing 99 percent accurate information about Down syndrome status, instead of the 87 percent accuracy of more traditional tests.

In the near term, when such tests penetrate the U.S. market, no doubt the same parents who spend years prepping their children for preschool interviews will buy these genetic tests. They will spend a lot of money; some will delay having children as they wait for "better" DNA predictions to come online. Others will hire companies to extract their sperm and eggs to create fertilizations that display better polygenic scores. It will be a very lucrative situation for teams of geneticists, lab personnel, and genetic counselors, but perhaps of dubious value for nervous parents. Perhaps soon-to-be parents will include in their pregnancy announcement posts on Facebook the hair color, (predicted) longevity, and risk for heart disease their eight-week-old fetus will have, in addition to the boring pink or blue stereotypical indicators of child sex. Then the parents will wait anxiously for the results over the next decades as the children are born and grow up, frustrated if the predicted phenotypes do not emerge as promised.

Some of these ideas are currently being tests in real time. BGI (formerly the Beijing Genomics Institute) is studying the genomes of about 2,000 highly gifted individuals in the hopes of better understanding the genetic etiology of great intelligence. As the Chinese government was supporting this endeavor, some people, like the University of New Mexico evolutionary psychologist Geoffrey Miller, suggested that the results could be used to test embryos in order to identify the "smartest" ones. In addition to these *Gattaca*-like scenarios,[8] there are concerns that genetic data can be used for discrimination. There is also the fear that links between diseases—along with traits like cognitive ability, and genetic variants that are more prevalent among some racial groups—may fuel discrimination.[9]

A more subtle, useful, and inequality-increasing implication of better genetic testing, though, will be what potential parents do with information about fetuses' *APOE* (a gene related to Alzheimer's disease) and *BRCA1/2* (genes related to breast cancer) status. Over time, embryonic or fetal selection, if not broadly covered by insurance, could shift the populations of those who eventually get breast cancer or dementia to be more strongly related to parental resources—a small step toward genetic stratification. This also means that specific genetic illnesses will increasingly be indicative of socioeconomic status, potentially shifting resources away from places like the Susan G. Komen for the Cure breast cancer organization. More generally, the typical "diseases of poverty" could shift toward genetic diseases.

There are also implications for the distribution of traits apart from diseases. Will parents use the information from candidate genes or polygenic scores for sexual orientation? Currently, there are no replicated genetic predictors of sexual orientation, but that does not mean this will always be the case. After all, many classic heritability studies suggest that the genetic component of sexual identity is probably around 50 percent.[10] What about skin color, which appears to matter to the life chances of both blacks and whites in the United States and elsewhere and is largely a product of our genetic code?[11]

As we transition to a situation of full testing, we will also need to consider those individuals caught in the middle of the technological arc—early enough to know they have an increased risk of an illness but too early to have a cure for the illness. Many people will learn their gene-carrier status at younger and younger ages. An important question is if, how, and when to provide the information and how we will collectively respond to bad news. For example, a study showed that providing information of their carrier status to people at risk for Huntington's disease cut the likelihood of getting a college degree in half for those who received bad news.[12] After all, why invest in the future if it will be cut short? But there is, as we will discuss below, hope, because there is new technology that can fix genetic mutations that cause diseases like Huntington's. But the larger issue of how people and their families will respond to receiving information about polygenic risk scores and what they mean is, itself, unclear.

In addition to concerns about prenatal DNA testing, there are fast-approaching implications of post-natal DNA testing. In September 2013 the U.S. National Institutes of Health (NIH) announced a $25 million pilot program to sequence the genomes of newborn children for the purpose of medical screening. Alan E. Guttmacher, director of one of the agencies overseeing the program, stated, "One can imagine a day when every newborn will have their genome sequenced at birth, and it would become a part of the electronic health record that could be used throughout the rest of the child's life both to think about better prevention but also to be more alert to early clinical manifestations of a disease."[13]

The potential merits of this type of program are obvious: they could detect genetic diseases more effectively than traditional forms of screening in the hope of successfully treating them. This information may also be helpful in identifying those vulnerable to other conditions of interest to social scientists. For example, genetic information may be predictive of learning disabilities, thus allowing us to intervene with specialized educational programs earlier than is presently possible before symptoms of say, autism or dyslexia, even manifest.

Along with this promise, there are several potential dangers associated with this type of program. Who has the rights to this genetic information? Newborns cannot give consent to have their genome sequenced, but they must live with the consequences of this decision for the rest of their lives. Should information about adult-onset conditions such as depression or high blood pressure be disclosed to parents, effectively taking away a child's right not to know (in the future)?

It remains unclear how parents will respond to being given detailed information about their child's genome. A major concern is that, as with most medical screening tools, genetic screening will lead to many false-positive results. This means that a child may never be stricken with a disease but the family must live under a cloud of fear based on its possible emergence. Parents may also get additional information about characteristics like personality traits and cognitive ability that may alter how they rear their children.

As we have described earlier, the relationship between genes and most disease, as well as other life outcomes, is probabilistic rather than deterministic. A relatively small number of diseases are like Huntington's in that a version of a single gene determines one's fate. Most diseases are the product of both genetic and environmental factors. This makes it hard to use the information from a genetic test, especially one that is administered to newborns. The research teams participating in the NIH pilot study do not plan to give parents all of the genetic sequence data. And yet, given the falling cost of sequencing, some parents will nonetheless soon have access to this type of information.

GENETICS INFILTRATES RELATIONSHIP SELECTION

And what will happen to those Alzheimer's-risky *APOE4* carriers (who will become rarer in the population and more likely to come from low socioeconomic status families) as they become teenagers and young adults and want to date and get married? How long will it take before dating services merge with genetic services and promise to both screen out *APOE4* carriers and provide a polygenic score for the future earnings (or longevity, or fecundity) of potential mates? eHarmony compiles data from a four-hundred-question survey in order to suggest matches between clients,[14] but a next step would be to add another million or so pieces of (genetic) information from each person to help inform the matches. Clients could match their preferences for (current) athletic (or slim) body build against genetic predictions of future BMI or future health expenses of a potential partner. Or a new, sophisticated generation of gold diggers could try to find partners with high levels of wealth and low predicted longevity.

New information on potential partners will usher in a new set of decision making trade-offs that are currently unavailable. We will be confronted by different aspects of "trading up"[15]—choosing matches who are currently beautiful but full of (genetically predicted) health issues and frailty later. Some might consider physically unattractive matches if they have high "genetic potential" that can be passed on to the kids—the next step after perusing binders full of Ivy League student-athletes in sperm banks.[16]

This process may lead to a different sort of stratification and separation of the matching market—into a (phenotypically stratified) dating market and a (genotypically stratified) marriage market. Of course, these two markets operate now. People can choose smokers with a self-reported interest in "having fun" for short-term or dating matches and then, later in their lives, choose partners with no bad habits and a lot of profile information devoted to career interests (or children) for long-term partnerships.[17] Will *APOE4* carriers be sorted out of the long-term match service sites—not formally denied services through a rule by eHarmony but informally shunned by other users? This extreme situation is unlikely, but some theories of matching markets would suggest that *APOE4* carriers will be disadvantaged in matches and will need to "settle" for partners that they would not have chosen before the genomic revolution revealed their status. And how will people with (genetically predicted) low fecundity do?

But there is another potential wrinkle: with genetic information from both partners (not just the information in the binders), dating services may attempt to predict which *combinations* of genetic partnerships might produce advantaged offspring through gene-gene interactions. Recall the importance of the genetic card-deck shuffle that happens after fertilization—we pass on only about half of our advantaged or disadvantaged genetic endowment when we have a child, gaining the other half from our partner. But there could be partners who are *particularly complementary* with your own DNA but whom you would not have suspected just by leafing through the sperm-bank binder or completing the four hundred survey questions on eHarmony or OkCupid. To the extent that matching services become adept at suggesting these combinations, what will become of other ways of finding a long-term partner? While we did not find evidence of increasing genetic assortative mating among those born in the twentieth century,[18] those results came from birth cohorts who had been blindly dating—genotypically speaking, that is. It is likely that once genotypic information becomes readily available, some persons will make use of it. And if it follows the patterns of most medical technologies,[19] it will be high-status individuals who make use of these data first and most often,[20] leading to a stratification in the level of genetic assortative mating. Even if the predictive power of scores is low, sorting on them may follow a class gradient and induce its own second-order effects on inequality.

MOVING TOWARD PERSONALIZED ENVIRONMENTS AND POLICIES

If some dating sites might suggest matches based on genetic complementarity, what other types of matches might we think about? In particular, we will consider matches based on complementary features of genetics and *environments*. This is an extension of the gene-environment interaction discussion in chapter 7, which concerned matching people with drug therapies that work best for their genotype—but now we raise the stakes.

When—if ever—should we act on evidence of important gene-environment interactions? An extreme example would entail removing dandelions from highly enriched environments (classrooms with extra teachers, for instance), since they are relatively unresponsive to such settings. (Indeed, to the extent that we recognize disadvantaged or challenging environments, we should place dandelion kids there, so the argument would go.) However, as we discussed in chapter 7, the current ability to predict dandelion status is very crude. But over time, one might imagine the technology and data expanding enough that we move from the question of whether we *can* accurately place students into milieus that match their genotypes to the question of whether we *should*. The efficiency arguments are straightforward—we should not waste resources on those who are unaffected by them (such as providing medical treatments for healthy individuals). The equity implications are more complicated, however, because dandelion status is probably outcome-specific—children could be dandelions with respect to educational enrichments but orchids with respect to athletic enrichments, or, more narrowly, could be dandelions with respect to their mathematical development but orchids for reading comprehension. The equity issues, then, may focus on the existence and availability of enriched environments across phenotypes. If we focus attention on development-enriched environments for math, then the math-orchids can blossom (the math-dandelions are unaffected), but the reading-orchids (and the athletic-orchids) lose ground, from a relative standpoint.

For an example of how genetics can create inequalities in response to policies, take the case of the Vietnam era draft and the GI Bill. We discussed how deploying the natural experiment of the draft lotteries of the late 1960s solves the problem of causal inference in assessing the impact of military service during the Vietnam War since draft numbers were truly randomly assigned, as if they were drugs and placebos in a medical trial. We showed how being drafted had a differential impact on life-long smoking behavior and, by extension, cancer risk, depending on one's smoking genotype. So by randomly exposing GIs to tobacco and stress, the draft not only created health inequalities between veterans and nonveterans, it also created inequalities that would not have otherwise existed within the veteran population. That is, in a world of no tobacco exposure, the smoking genotype does not matter—nobody smokes. But throw in a random exposure to tobacco, and genotypic inequality rears its head. Of course, the genetic heterogeneity in response to the "treatment" of tobacco exposure and the stress of war is happening whether we measure it or not.[21] A remaining question is, armed with knowledge of the gene-environment interaction between stress and genotype, should "we" (i.e., the government) intervene in order to "personalize" policy, which in this case might suggest focusing recruitment efforts on specific genotypes?

It is not only negative exposures that can generate genetically filtered inequalities. Positive policies meant to give everyone a leg up can also generate inequalities based on genotype. Take the GI Bill, considered one of the greatest policy triumphs of the second half of the twentieth century. The GI Bill, along with Pell Grants and other policies, helped bring college attendance within reach for a huge number of Americans who otherwise would have been shut out of higher education and the career opportunities that follow. Research by the economist Jere Behrman and colleagues has shown that the GI Bill, in fact, reduced the influence of family (i.e., class) background on educational attainment.[22] So, thanks to the GI Bill, those who were drafted during the Vietnam War era do show an average higher educational attainment than those who were not drafted. And of course, since many young men and women who cannot afford college enlist in the military, partly because of the GI Bill, the policy has an average effect of reducing class disparities in college attendance. However, we found that those veterans who were most genotypically prone to pursue higher education were the ones, when drafted, who benefitted from the GI Bill by obtaining more education. Those who had a lower educational genotype were not as likely to avail themselves of this opportunity. Thus, the GI Bill could have generated inequality within the veteran population even as it provided opportunities to veterans who might not have otherwise had them.[23] As with the smoking example, these inequalities were emerging whether we measured them or not. Genotyping individuals and observing that genotype interacts with the policy merely sheds light on an otherwise hidden form of stratification. But genotyping also allows the question, should we further target college attendance policies toward economically disadvantaged people with more favorable polygenetic scores?

This sort of prismatic effect is to be expected whenever opportunity is presented but not "enforced"—those genetically (or socially) most capable of taking advantage of government-provided (or other) opportunities are the ones who benefit. As some dimensions of stratification are reduced, others are magnified. Contrast this with a policy that enforces school attendance for everyone rather than just providing the chance for all. Sounds draconian, but the United States has a long history of ratcheting up the number of years of compulsory schooling, from essentially zero about 150 years ago to about ten or eleven years in most states today (depending on the birthday of the student).

As we have repeatedly mentioned, genetically differentiated responses to universal (or targeted) policies exist whether or not we measure individual genotypes. But purposefully using polygenic scores (like the education score in the Vietnam study) to better sort people into environments is an entirely different matter. First, aside from the ethical issues, the cause-and-effect structure underlying these scores is not understood. Why someone with a high polygenic score for schooling will attain more schooling is not known, and more pertinently, how these scores may or may not interact with different aspects of the environment is even murkier at the present time. Indeed, a specific feature of the polygenic scores is critical here. Since the scores are calculated using data sets that combine populations from all over the world,[24] who are exposed to vast differences in environments, the information gathered from these scores is precisely the genetic signal that is likely to be the *most invariant* under routine environmental differences and exposures. In creating these scores, we are capturing mostly the genetic variants that work the same way in every environment in the Western, industrialized world rather than the genes most sensitive to environmental differences. This means that sorting people into environments based on polygenic scores may be *precisely the wrong way* to use the scores.

That said, as mentioned in chapter 7, there are new efforts underway to generate scores that predict not average levels of a given phenotype, like height or education, but rather those that correlate with *variation* in an outcome. These so-called plasticity scores could identify individuals who are more or less responsive to environmental effects. For example, in 2012, a paper in *Nature* used the vGWAS approach to identify loci that were associated with variation in height and BMI.[25]

THE PUBLIC AVAILABILITY OF SENSITIVE GENETIC INFORMATION

Once we are all in happy relationships producing superkids with unique combinations of DNA from each partner that maximizes children's outcomes, what happens when hackers infiltrate 23andme and capture our genetic data? We need to prepare for a situation in which we have *public* information about everyone's genetics. Our addresses, medications, professional accomplishments and failures, family photos, and other formerly private information is becoming easier to find online. Employers and colleges look at Facebook and Twitter when considering candidates (as do potential dates). With large-scale hacking attacks that release tens of millions of people's private information all at once, the likelihood that your genetic profile will be hacked in the future is high. Indeed, a group of researchers were recently able to identify anonymous individuals participating in a study via their published genetic information, Google searches, and a genealogy website.[26] Even more troubling, the researchers were able to identify additional members of participants' families even though they were not part of the study. Genetic information is increasingly being used to track down long-lost relatives, especially among adopted children searching for their biological parents. There are also similar stories of searches like this for sperm donors. What, if anything, should government have to say about these and other uses of genetic information by individuals, corporations, and the state?

Legislatures have only recently begun to enact legal measures against genetic discrimination—but these laws have primarily focused on insurance companies and employers. What about broader genetic discrimination and the ensuing genetic stratification that may result from this practice? People who have "good" genetics may start broadcasting this fact, and anyone who does *not* proclaim their polygenic score values will be suspected of having inferior scores. Or will these scores remain akin to ACT or SAT scores, something potentially knowable but not discussed in polite conversation?

Some social consequences of our genotyped future can already be felt now. For example, in many jurisdictions when someone is taken into custody by law enforcement, a sample of DNA is taken. This DNA is then available to compare with all other criminal cases that contain DNA data. Many of the stories in the media are about DNA exonerating—often years later—wrongly convicted individuals. The Innocence Project—started in 1992 at Cardozo Law School—pioneered the movement to initiate appeals of flimsy convictions based on often-overlooked biological samples. And the success of the Innocence Project—along with its heart-wrenching tales of wrongful imprisonment—seems to suggest that civil libertarians should welcome this new era of forensic science.

On the prosecution's side, DNA has been a boon (except in the case of O. J. Simpson), when it has been available at a crime scene or obtained from a rape kit. Now suspects can be definitively matched to the scene of the crime via blood or semen, for instance. At first blush, then, it seems like DNA in the courtroom is an unalloyed good: it reduces the errors in an otherwise faulty system that relies on fallible human testimony and other less "scientific" approaches to establishing perpetrator identities.[27]

But like most technologies, forensic genetics has a tendency to reproduce existing inequalities. If you are caught because you have a prior conviction and your bodily fluids at the scene of a new crime match those that are on file, that is unfortunate for you, but it is hard to make a case that it is inherently or systematically unfair. People who have committed prior crimes are more likely to be caught than those who are not in the system. However, if your brother or mother has been genotyped by law enforcement—even if you have led a squeaky clean life until now—then you are also more likely to be caught up by your genetic fingerprint. That is, your DNA will identify you as a first-degree relative of your sibling. And that information, plus a little detective work, is almost as good as you, yourself, appearing in the data set. DNA fingerprinting can even identify cousins or grandchildren—although with less certainty. So if your relatives are more likely to have been registered in the database, you are more likely to be located, even if you yourself have no priors. Of course, in this case, the unfairness does not lie in the fact that you were tripped up by your DNA but rather that the person who committed a crime and who comes from a more advantaged background, and thus does not have relatives in the database, gets away with murder (literally or figuratively). Add in the sort of class or race stratification that is thought to exist in the criminal justice system, and you have a perfect storm by which DNA amplifies existing inequalities.

FITTER, HAPPIER . . .

As frightening as it is to have your genetic information publicly posted on the Internet (or in a criminal database), one difference between public records of sensitive financial information and sensitive genetic information is the following: you can change your credit card account number, but not your genes.

Or can you?

Actually, we may soon be able to change our genetic code. New technology, leveraging the CRISPR/Cas9 system ("CRISPR," pronounced "crisper," stands for "clustered regularly interspaced short palindromic repeats," and "Cas9" is shorthand for "CRISPR-associated-9") is inching us toward the brave new world of editing our genes. In short, the technology deploys segments of viral DNA that have been previously incorporated into the genome of a bacterium to edit specific sites on the genome. These *cas* genes encode enzymes (Cas9 endonucleases) that can cut one or both strands of the DNA at specified points, excising a small section in the process (the part one would like to remove or replace). Donor DNA (supplied by the scientist) is then inserted as the strand heals itself. CRISPR was first used in 2012 and has since been deployed in yeast, flies—and humans (so far, in the form of nonviable embryos).[28] Are you unhappy with the *APOE4* variant your parents bequeathed you? Change it!

Clearly this is an explosive technology, with implications complementary to and on par with the sequencing of the human genetic code itself. There is now a large effort to enforce a moratorium on any CRISPR applications in humans because the science is so far ahead of the thoughtful consideration of its social and ethical implications.[29] Unfortunately, such an effort to ban human applications works only if all countries agree, and the People's Republic of China has not yet signed on.

In April 2015, scientists from China published results describing their attempts to alter nonviable human embryos to correct a mutation in the DNA that causes an often lethal blood disorder called beta thalassemia (the larger class of beta-thalassemia mutations affects about 1 in 100,000 people).[30] Although the moratorium partially "worked" in that the leading scientific journals (*Nature* and *Science*) decided against publishing the findings, the moratorium failed in its main objective of hitting the pause button on the use of this technology in humans. It turns out that the Chinese efforts were largely for naught in the end, because the scientists either failed to edit the right part of the DNA or failed to edit the DNA at all. Indeed, eighty-five percent of attempts were unsuccessful. The four embryos that were successfully edited suffered from a mishmash of edited and unedited cells (called genetic mosaicism) and thus were not viable either. Although CRISPR is in its early days, we do need to plan for contingencies when it is more successfully used in humans.

Gene editing necessitates a deeper discussion about identity. Is someone the same person after editing his genome? If the person has an identical twin, does their twin status convert to fraternal following the gene edit? Are there gene edits that, biologically, suggest a reordering of familial relationships and status? For example, there are edits to the genome by which a traditional DNA paternity test would fail to recognize the relationship. Indeed, a DNA maternity test could also fail. If both your parents have a T at a given locus, and you edit your genome to have an A at that locus, part of your genome exists outside of the family tree.[31] As the technology develops, allowing a larger number of edits on a single genome, a larger separation will appear between how we view biological familial relationships and socially constructed familial relationships. Indeed, some genome edits could, genetically, move a person from one family tree to another. In addition to the potential of micro, family-based disruptions to identity, there are also possible macro-level disruptions. There is evidence that a single SNP determines blue eye color.[32] How would our understanding of race and ethnicity classifications in the United States incorporate an Asian American who chooses to edit this letter of his genome? Would someone with two African American parents and blue eyes be considered mixed-race? It is an arbitrary example, but one that also showcases both the foolishness of our racial/ethnic categorization scheme and also a way to destroy it—by upending our traditional view of links between phenotype, ancestry and racial categories.

There are more complex questions as well. In addition to your life's blueprint ("software"), your genome is a historical document. It provides information about who your forebearers were, your relationship status with everyone on the planet, and hints of the geographic path your family took to arrive in the present. Editing your genome changes this history; it makes genetic history irrelevant in a tangible way. In this way, at least conceptually, genome editing represents the possibility of a fresh start.

A different aspect to consider—as we project our imagination across centuries rather than years—is how gene editing may affect the species as a whole. On the one hand, we could envision a relatively rapid disappearance of many health disorders—both those caused by single genes and those with a more complicated genetic architecture. Birth defects could disappear, as could a number of cancers, many diseases of the brain (e.g., Alzheimer's), and other devastating illnesses. But not all conditions are so easily fixed. Some will presumably be too complex to attack through gene editing, even as the techniques improve. Other conditions may give us pause before eliminating them, even if the technology allows it. What about Asperger's syndrome? People with Asperger's often have both social deficiencies (as judged by others) but also great abilities. What are the costs to society of deciding to eliminate this condition?[33] Since CRISPR edits the germ line (so not just the current generation, but also future generations), it is conceivable that society could eliminate both Huntington's disease and also Asperger's syndrome in a single generation. If we believe the orchids-dandelions theory (discussed in chapter 7), then we know that orchid status is valuable at the individual level but potentially dangerous at the species level. Recall that dandelions, while having a lower than optimal fitness, help the species hedge against environmental changes, due to which all the orchids may be killed off. CRISPR technology jeopardizes this symbiotic arrangement—now, dandelions can become orchids. At the individual level, this conversion will make sense; however, a mass conversion places the species at risk. With no more dandelions, we would lose the ability to hedge our bets through a reduction in genetic diversity. In addition to these worrisome long-term implications, we should also consider the ways the CRISPR technology could impact the near term.

Of course, as with DNA fingerprinting, prenatal genetic selection, or even DNA dating apps, it is likely that those with resources will be the first to take advantage of such a novel technology. This augurs a world in which the separation between "natural" and "political" (i.e., social) inequality collapses completely. The distinction between the natural and the social is not only made by behavioral geneticists who seek to parcel out the effects of nature and nurture, but also dates at least as far back as Jean-Jacques Rousseau, who claimed in *A Dissertation on the Origin and Foundation of the Inequality of Mankind* that

> I conceive that there are two kinds of inequality among the human species; one which I call natural or physical, because it is established by nature, and consists in a difference of age, health, bodily strength, and the qualities of the mind or of the soul; and another, which may be called moral or political inequality, because it depends on a kind of convençal inequality, and is established or at least authorized by the consent of men. This latter consists of the

different privileges, which some men enjoy to the prejudice of others; such as that of being more rich, more honoured, more powerful, or even in a position to exact obedience.[34]

Already, of course, families who are wealthy can correct children's myopia through laser eye surgery, afford healthier food and a safer environment for their children, and so on. If a father enhances his eyesight or learns a foreign language or takes vitamin supplements, he may alter what was once his natural state by means of his social position; however, he does not necessarily pass those advantages on to his offspring (except through creating a healthier family environment and culture). In order to bestow the advantages on the next generation, the father needs to speak that foreign language to his infant, or feed her vitamins, or pay for eye surgery when she gets old enough. But germ-line genetic alteration is qualitatively different. In the case of gene editing (or even genetic selection of embryos), a father can convert any form of financial, human, or social capital he possesses into natural capital for not only his children, but for their children's children as well. The levee between the social and the natural forms of inequality will have been completely breached.[35] What the ensuing flood will wash up is anyone's guess, but it is coming.

EPILOGUE: GENOTOCRACY RISING, 2117

The not-so-young parents sat in the medical office laden with plastic models of embryos at different stages of development and walls covered with cladograms of social stratification. This was not their ob-gyn's or pediatrician's office but rather their reprogeneticist, a medical subspecialty that emerged in the late 2050s, with at least one practitioner in every high-end fertility clinic. In the parents' latest round of in vitro fertilization, thirty-two viable embryos had been created, and so now they faced a choice that had become fairly common. Anxiously, they pored over the scores they had received from the clinic for various traits. Half of the blastulae were fairly easy to eliminate because they had higher-than-average risks for either cardiovascular problems or schizophrenia or both. That left 16 potential babies from which to choose. Ten were girls, and since this was their second child, they wanted a boy to complement their darling Rita, now entering the terrible twos. Of the male embryos, one was predicted to be significantly shorter than the parents and his older sibling. Yet another had a greater than a one-in-four chance of being infertile. And since this was likely to be their last child due to advanc-

This epilogue adapted from D. Conley, "What if Tinder showed your IQ? A report from a future where genetic engineering has sabotaged society," *Nautilus Magazine*, September 24, 2015, http://nautil.us/issue/28/2050/what-if-tinder-showed-your-iq.

ing age, they wanted to maximize the chances that they would some-day enjoy grandchildren.[1]

That left four male embryos. They scored almost identically on disease risks, height, and body mass index. Where they differed was in the realm of brain development. The brightest one scored a predicted IQ of 150 and the worst a "mere" 130. A few generations earlier, an IQ of 130 would have been high enough to ensure an economically secure life in a number of occupations, but with the advent of voluntary artificial selection, a score of 130 was now only slightly above average (since the tests had not been renormed). By the mid-2070s, it took a score of 140 or higher to insure your little one would grow up to become a knowledge leader. Extreme beauty, sports ability, or tremendous skill as a performer were even more of a long shot as a path to economic security than they had been in the early 2000s.

But there was a catch. There was always a catch. The science of reprogenetics—self-chosen, self-directed eugenics—had progressed over the years, but it still could not escape the reality of evolutionary trade-offs, such as the increased likelihood of disease when one maximized on a particular trait and ignored others. Or the social trade-offs—the high-risk, high-reward economy for reprogenetic individuals—in which a couple of IQ points could make all the difference between success or failure, but where stretching genetic potential to achieve those cognitive heights might lead to a collapse in noncognitive skills, such as impulse control or empathy.

Against this backdrop, the embryo predicted to have the highest IQ also had a significantly greater chance of being severely myopic to the point of uncorrectable blindness—every parent's worst nightmare. The fact that this genetic relationship between intelligence and focal length had been known about—or at least suspected—for decades (since a 1988 paper in *Human Genetics*) did not seem to dampen the mania for maximizing IQ during the last few decades.[2] Neither did the fact that the correlation worked through genes that controlled eye and brain size, leading to some very odd-looking, high-IQ kids.[3] (And of course, anecdotally, the correlation between glasses and IQ has been the stuff of jokes for as long as ground lenses have existed.)

The early proponents of reprogenetics failed to consider the basic genetic force of pleiotropy: that the same genes have not one phenotypic effect but multiple ones. Greater genetic potential for height also meant a higher risk score for cardiovascular disease. Cancer risk and Alzheimer's probability were inversely proportionate—and not only because if one killed you, you were probably spared the other, but also because a vigorous ability to regenerate cells (read: neurons) also meant that one's cells were more poised to reproduce out of control (read: cancer).[4] And as generations of poets and painters could have attested, the genome score for creativity was highly correlated with that for major depression.

But nowhere was the correlation among predictive scores more powerful—and perhaps in hindsight none should have been more obvious—than the strong relationship between IQ and Asperger's risk. Each additional 10 points over 130 also meant a doubling in the risk of being on the spectrum. And since the predictive power of genotyping had improved so dramatically, the environmental component to outcomes had withered in a reflexive loop. While in 2017, IQ, for example, was, on average, only two-thirds genetic and one-third environmental in origin by young adulthood (according to work by psychologist Richard Plomin and others),[5] the fact that we started measuring the genetic component became a self-fulfilling prophecy. That is, only children with high IQ genotypes were admitted to the best schools, regardless of their test scores. (It was generally assumed that actual IQ was measured with massive error early in life anyway, so genes were a much better proxy for ultimate, adult cognitive functioning.) This prebirth tracking meant that environmental inputs such as early exposure to a wide vocabulary, which were of course still necessary, were themselves perfectly predicted by the genetic distribution. Such sorting then resulted in a heritability of almost 100 percent for the traits most important to society—namely IQ and (lack of) ADHD (thanks to the need to focus for long periods of time on intellectually demanding, creative work in a world in which machines were taking care of most other tasks). (Interestingly, low IQ is also a risk factor for both [low-functioning] autism spectrum disorders and ADHD.)

Who can say when this form of prenatal tracking started? Back in 2013, a paper in *Science* constructed a polygenic score to predict education.[6] At first, that paper, despite its prominent publication venue, did not attract much attention. That was fine with the authors, who were quite happy to avoid media glare for generating a single number based on someone's DNA that was correlated—albeit only weakly—not only with how far they would go in school, but also with associated phenotypes (outcomes) like cognitive ability (the euphemism for IQ still in use during the early 2000s). That said, from a scientific perspective, the *Science* paper on education was not earth-shattering because polygenic scores had already been constructed for many other less controversial phenotypes: height and BMI, birth weight, diabetes, cardiovascular disease, schizophrenia, Alzheimer's, and smoking behavior—to name just some of the major ones. Furthermore, muting the immediate impact of the score's construction was the fact that—at first—it only predicted approximately 3 percent of the variation in years of schooling or IQ. Three percent was less than one-twentieth of the variation in the bell curve of intelligence that was reasonably thought to be of genetic origin.

Instead of setting off a stampede to fertility clinics to thaw and test embryos, the lower predictive power of the scores in the first couple of decades of the century initiated a scientific quest to find the "missing" heritability—that is, the genetic dark matter where the other (estimated) 37 percent of the genetic effect on education was (or the unmeasured portion of IQ's genetic basis). With larger samples of respondents and better measurement of genetic variants via genotyping chips that were improving rapidly, the dark-horse theories for missing heritability (such as Lamarck's epigenetic transmission of environmental shocks) were soon slain, and the amount of genetic dark matter quickly dwindled to nothing.

At first, clinicians and the wider public hardly noticed. They had, at the time, been enthralled by the CRISPR/Cas9 technology, which was lighting up the press with talk of soon-to-be-awarded Nobel prizes. As the twenty-first century rolled on, the gene-editing system had a huge impact on human disease and well-being: it not only effectively eliminated all single-gene birth defects from the human population of the developed world, it also turned cancer into a chronic (if still painful) condition that could be treated by genetic alterations to excise oncogenes gone haywire. It also improved the food yields and nutritional value of staple crops once the political opposition to genetically modified organisms was overcome.

But once parents started "enhancing" their germ lines by editing their ova and sperm cells, the science hit a political wall. Initially, the changes that these intrepid reprogenetic practitioners were evincing were relatively harmless: switching from brown eyes to blue, or from attached earlobes to detached, or from dark hair to light. That, in fact, was the limit of what could be done by editing a single gene or a small number of genes, since most dimensions along which humans varied—everything from height to extraversion to metabolism—were highly polygenic—resulting from the sum total of thousands of small effects spread across the twenty-three pairs of human chromosomes.

A breakthrough came when the amount of DNA that technicians were able to reliably amplify from an early-stage embryo reached a level that allowed for full-genome sequencing. Rather than merely having a bird's eye view of the chromosomes to look for major damage, such as duplications or deletions, now each base pair could be examined in turn. Once the code for the embryos had been unlocked, in 2022, it was a simple matter of running the results through a spreadsheet to identify predicted phenotypes.

The practice quickly spread down the socioeconomic ladder as employees demanded that their health insurance pay for this sort of screening; eventually such coverage was mandated by law. This change in health policy was initially motivated with the goal of reducing obesity and depression (which were covered conditions in health insurance policies), but eventually the practice spread to nonhealth and noncovered phenotypes, such as IQ and even impulse control.

At the same time, the merger between 23andme—the largest genetics database in the world—and InterActiveCorp (owner of Tinder and OkCupid), and their subsequent integration with Facebook, meant that not only were embryos being selected for implantation based on their future abilities and deficits, potential spouses were also sorting on the basis of genotype. Rather than just screening for nonsmokers, why not screen for nonsmokers who are also genotypically likely to pass that trait on to one's potential offspring?

Of course, just as there were always people who refused to sign up for Facebook, the naturalist, alternative subcultures—those who mated "blindly"—also grew and thrived. Little did these subaltern "geno-resisters" know that their DNA was surreptitiously being used by the reprogeneticists to test their statistical models and refine their polygenic scores. The simulations required the greater amount of variation that was provided by natural reproduction; as more and more parents practiced reprogenetics and minimized the presence of certain DNA variations and maximized that of others, there was little variation upon which to test for potential cross-gene interactions among the artificially reproducing population. Second, and more important, because the majority of the population now sorted and invested in their children based on their genetic scores, the effects of those scores within that population became circular and self-fulfilling, adding little to their underlying predictive power. To achieve greater efficiency through improved genotyping accuracy, one needed an environmental landscape that still followed its own logic that was orthogonal to genotype.

The social world soon bowed to this new auto-evolutionary reality. Not only did admissions testing for schools give way to genetic screening, but the educational system fragmented into stratified niches based on specific combinations of genetically based traits. There were programs for those who ranked high in athletic ability and were neurotypical, and others for those who ranked high both in motor skills and on the autism spectrum. There were jobs that required ADHD, and those that avoided it. All of this was done in the name of greater economic efficiency. But maximum IQ was always king in this world.

The result for families was the declining cohesion of anything that could be called a family or household unit. One might think that greater parental control over the genotypes of their offspring might result in families that specialized in a tendency or skill—say, visual-spatial or verbal ability—and created a domestic culture geared toward fostering it. The reality, however, was that sibling differences were accentuated to an extent never before seen because parents had to maximize what they could from a small sample of embryos rather than specialize in a particular niche.

Most sociogenetic consultants suggested to parents that they might want to maximize similarity to their older child to avoid this problem of how to parent two vastly different genotypes in the same household. Genetic differences, they explained, are actually magnified within families as parents consciously or unconsciously try to provide the investments and environments that each child needs to achieve their genetic potential. Small differences in IQ or athletic ability (that might have been obscured by environmental differences when comparing two kids from different backgrounds with the same genotypic score) become outsized in their effects when viewed against the control group of a sibling. And this becomes, like so many social dynamics, a self-fulfilling prophecy that leads to the ironic situation of greater differences within families than between them.

This familial dynamic could have been predicted by the early-twenty-first-century research studies, some of which had shown that the effect of the polygenic score for cognitive ability was stronger within families than between them. Genetic scores predicted the differences between siblings better than they predicted the differences between randomly selected individuals from different parents. A sibling with a one-standard-deviation better education score than his brother is likely to complete a half-year more schooling, on average. But compare two strangers with the same difference in their genetic score for education, and the average difference in schooling is only a third of a year.

Dramatic changes to socialization processes were not limited to the family unit. Recent history also witnessed the rise of highly specialized boarding schools that catered to younger and younger children each year. The interaction between a parent's genetic score and that of her offspring had the potential to magnify effects (for better or worse), so parents were increasingly influenced by slick marketing campaigns that suggested educational environments fine-tuned to a child's particular combination of genotypes. It was a severe and conscious case of an environment structured by genetics. Adding to the frenzy of finding the "right" placement—which made the competition for private preschool slots in places like Manhattan and San Francisco in the early 2000s look like open enrollment—was research that showed that not only did a parent's genotype blunt or accentuate the genotypic effects for a child, the entire genotypic environment mattered. How your genotype for behavior played out depended on the distribution of genotypes around you.

But unlike the tall poppy or the differently colored wildflower that attracted the pollinator and was advantaged in the game of reproduction by being different, the research showed positive genotypic peer effects. Being around those with your same predicted phenotype was good for you—there was strength in genetic numbers. The end result was micro sorting in schools and in almost all aspects of social life, including marriage and occupation. In 2000, those in the medical fields had the highest rates of in-marriage; for example, 30 percent of married male nurses were married to other nurses. By the 2040s, nurses married each other at a 90 percent rate, and it was hard to find any profession that was below 80 percent.

Micro sorting should not have been able to last, since the entire point of sexual reproduction, as evolutionary biologists had long ago told us, was to maintain genetic variation into the population. Compared with asexually reproducing species, mating species—which inefficiently pass on half as many genes as they would by merely cloning themselves—are more robust to genetic drift and environmental challenges (such as rapidly evolving parasites). Recombination during the production of sperm and eggs (meiosis) means that advantageous alleles can be pooled and deleterious ones purged from surviving offspring. In this way, sex speeds up selection. This, of course, was what made the entire reprogenetic endeavor so rapidly realized in the first place.

Back in the medical clinic, the sociogenetic consultant suggested to the not-so-young parents who wanted a boy that they pick the embryo the most similar to their daughter Rita, regardless of which one might be the most successful. If they were immunologically similar, they could interact without fear of literally killing each other. Alas, the consultant's advice was not heeded. The parents opted for the 150 IQ.

APPENDIX 1

WHAT WE TALK ABOUT WHEN WE TALK ABOUT MOLECULAR GENETICS

The central dogma of molecular biology is DNA → RNA → protein. DNA provides the blueprint, which, aside from de novo mutations (the sort that sometimes lead to cancer) or mosaicism (when, for example, some of the cells in an individual's body are of a different origin such as maternal or fraternal), is identical in every cell of the body. The human genome (which includes the DNA) is stored in the nucleus of each cell (with the exception of red blood cells, which have no nucleus) in 23 pairs of chromosomes as well as in the mitochondria (power plants) of each cell. Mitochondrial DNA (mtDNA) is inherited only from the mother, since it arises from the ovum (although there is some debate as to whether some mitochondria from the sperm cell penetrate the egg and survive in the development of the fertilized zygote). The nuclear DNA is inherited from both parents, one of each pair of the 22 autosomal chromosomes (those other than the sex chromosomes) coming from each progenitor. As for the sex chromosomes, under typical circumstances the mother always provides an X (female) chromosome. The father provides an X (making the offspring female) or a Y (making it male). Thus, analysis of mtDNA allows us to peer back through the enate line, while Y chromosome analysis allows for characterization of the agnate line.

All in all, if we unfurled the DNA strands from the 46 chromosomes and lined them up end to end, they would be six feet in length and contain 3 billion base pairs. There are four bases: adenine (A), thymine (T), guanine (G), and cytosine (C). They have specific complementarity so that the double helical phosphate backbones can be joined only by A with T or C with G. Among these pairings, there is variation in about one in a thousand (high-end estimates put this at four in a thousand) at these positions—yielding a figure of 3 million single-base differences and the commonly cited notion that we are 99.9 percent genetically identical. If we also consider another common form of difference—copy number variants (CNVs)—we are an estimated 99.5 percent similar. Other forms of variation include structural variation in chromosomes, such as insertions or deletions (indels). The figures about similarity are somewhat misleading since small differences can lead to huge phenotypic differences.

In regions of the genome that encode for messenger RNA (mRNA; which transmits the template for proteins to ribosomes, on which proteins are assembled), triplets of bases known as codons specify which amino acid is called for in the assemblage of the protein (which are chains of usually 100 or more amino acids strung together like beads). There are also codons for "start" and "stop." If there is a change of nucleotide in the third position in the codon (say from CTA to CTG), this is usually known as a silent, or synonymous, mutation because it does not change the amino acid called for and thus does not affect the protein's makeup (although it can affect the efficiency of production). A change to either of the first two nucleotides in the triplet is called nonsynonymous and leads to a structural change such as an amino acid substitution (missense) or to stopping transcription (nonsense mutation).

The term "gene" generally refers to a protein-coding stretch of DNA, including not only the part that gets transcribed but also the promoter region (the part before the start of the coding region and where the transcriptome attaches to begin its work) and other regulatory regions known as enhancers (these are commonly found within the first intron but sometimes are present thousands of bases away). After an mRNA has been transcribed from DNA, it is edited by biochemical machinery that snips out introns and leaves exons, which will be translated into proteins. Further regulation of mRNA translation into protein (and even control over where the mRNA goes) is affected posttranscriptionally by the 3′ (pronounced "three prime") UTR (untranslated region) of the mRNA, which comes after the last amino-acid coding codon.

There are only about 20,000 or so genes in the human genome, each of which can produce about three different proteins on average (by the alternate splicing or pruning of introns). That figure is far below what most geneticists had expected for humans (for example, rice, which one would like to think of as a less complex organism than ourselves, has about 46,000 genes). This finding is noteworthy because it reveals the importance of gene regulation. That is, since every cell contains the same genetic blueprint, the differences between a neuron and a hepatocyte and an epithelial cell are all about which genes are expressed when. Likewise, differences among humans are not largely due to different protein structures but instead to the fine-tuning of gene expression at critical points in development. This realization goes hand in hand with the recognition that much of the non-protein-coding part of the genome is hardly "junk" DNA (which it has been called in the past) but instead is critical to conducting this symphony. For example, a form of RNA called micro-RNA (miRNA), which often binds to the 3′ UTR region of RNAs, actually serves an important role in regulating the process of translation. Other areas of the genome produce not full proteins but instead peptides, which are short strings of amino acids and can form a certain class of hormones as well as some neurotransmitters, such as endorphins (endogenous opioids).

Variation in gene expression is controlled by a number of factors, some of which are collectively termed epigenetics. Epigenetics has become a field of great excitement within the social sciences, possibly because of the notion that it reverses the causal arrow of traditional genetic analysis (from genome to phenotype), pointing it in a direction in which sociologists, for one, feel much more comfortable: from environment to genome. While traditional genetic analysis of behavior examines variation in the nucleotides that are fixed at conception and have effects that ripple out across life's course, social epigenetics examines how the environment affects gene expression through processes such as histone acetylation (addition of an acetyl

[COCH$_3$] group to one of the proteins [histone] around which DNA is coiled when stored) and DNA methylation (the addition of a methyl [CH$_3$] group to a CG sequence) that influence whether or not a particular gene gets turned on or off in a given tissue at a given time. Some scholars are particularly excited by the notion that such environmentally sensitive epigenetic marks may, in fact, be inherited across generations. If so, it would suggest that part of biological inheritance has environmental roots—that social factors such as wealth and poverty, incarceration, slavery, family processes, and the like can all be incorporated into the genome. It should be noted, however, that while intergenerational associations have been shown in, for example, DNA methylation patterns, other mechanisms have not been ruled out. At the same time, experimental evidence from animals is providing the basis for theories that some epigenetic marks that are conditioned by stimuli may, in fact, survive in offspring. The evidentiary bar for transgenerational epigenetic "memory" is rightly set very high since the current thinking is that the vast majority (if not all) epigenetic marks are erased during reproduction (meiosis—i.e., the production of a haploid gamete) in order to produce an omnipotent stem cell capable of becoming all cell types in the developing embryo (whereas epigenetic marks tend to constrain pathways of development). Meanwhile, there are many other pathways in addition to epigenetic marks by which information about the environment can be transmitted to offspring. Transgenerational epigenetics promises to be an exciting field for social scientists to watch in the next decade or two, regardless of whether it proves to be a revolution in our understanding of heredity and the nature-nurture dichotomy. At the very least, molecular biologists have complicated their own central dogma and now recognize many ways in which causal arrows point forward and backward and loop around the DNA-RNA-protein nexus. Social scientists ignore this epigenetics revolution at their peril if they seek a complete understanding of human behavior.

APPENDIX 2

A SECOND TRY AT REDUCING HERITABILITY ESTIMATES: USING GCTA AND PC METHODS

To run the heritability analysis on unrelated individuals (known as GCTA or GREML[1]), not only are pairs of individuals who are cryptically related not included, but the first few principal components (PCs) are purged from the data first so that any remaining allelic similarity between pairs of subjects is (it is assumed) the result of chance. Then we can be sure that environmental differences are not clouding the inference that we are detecting only genetic influences with this method. (Of course, it is worth restating that environmental differences that are caused by genetic differences are merely pathways—or endophenotypes, in the field's parlance—that form part of the way that genes have their effects.)

But what if there were still environmental differences that were confounding the estimates? We looked at putatively environmental measures over which individuals had no control—whether someone grew up in a rural or urban setting and the educational level of their parents—and found that when using this method for unrelated individuals, these measures were themselves heritable. Now it could be that for genetic reasons some people like country living and other people prefer city life. And because children share 50 percent of their genomes with their parents, we may have been merely picking up genetic influences on the parents' residential choices, but only weakly since the genetic signal was diluted through the offspring. But since we were looking at pairs of offspring, the effect should be doubly diluted, to approximately one-quarter of what the actual heritability should be if we measured the parental genomes themselves—assuming random mating.[2] Instead, we found a 30 percent heritability for rurality/urbanity. We could drive it down to about 15 percent by factoring out 25 PCs rather than the standard two to five, but we could not eliminate it completely. This lower bound of 15 percent would imply that for the parents the heritability is minimally 60 percent. This number is highly improbable. (This would be getting closer to the heritability for more proximate biological outcomes, such as height, which is approximately 80 percent.) We obtained similarly improbable estimates for maternal education. Thus, we thought something was fundamentally wrong with the GCTA approach. Perhaps environment was sneaking into the genetic side after all, just as we thought it might be with the twin models.

This suspicion was confirmed when we looked at the original *Nature Genetics* paper that pioneered the method using schizophrenia as an outcome.[3] The authors had performed a different robustness check than the one we tried above, and in fact they did something cleverer: they looked at each chromosome one at a time. If it was truly random variation due to genetic card shuffling and not tribalism (and thus differential environments) that was generating the estimates, then the genetic similarity of two individuals for one pair of chromosomes—for example, chromosome 4—should be unrelated to that pair's similarity on another set of chromosomes—for example, chromosome 12. The reason is that rather than shuffling and splitting one big deck, our genome is packed into 23 pairs of small decks. How one is shuffled should have no predictive power on how another is shuffled, unless there is population stratification of some kind, and deep common ancestry was causing some people to be more genetically alike than others. That is, if genetic relatedness overall was also driven by population structure (common ancestry), then we would see this ancestral signal in the fact that people who were more similar in chromosome 8 would also be systematically more similar in chromosome 4 (and 14, and 16, and X, and so on). There would be no other way to explain this pattern than to say it was not random shuffles that made people more similar *across decks* (chromosomes).

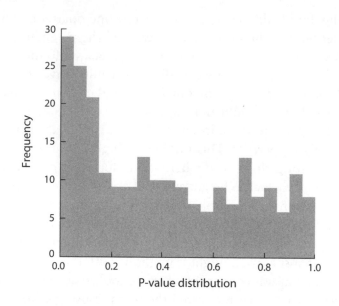

Figure A2.1. If it really were only from the randomness of recombination that some people are more related than others, we should see that chromosomes are uncorrelated in their degree of relatedness since they segregate separately. Instead, we see a shift to the left in the probability distribution of correlations between people across chromosomes. Chromosomes tend to cluster in terms of relatedness. This observation indicates that population structure (i.e., ancestry) is driving part of the overall relatedness distribution, thus suggesting that environmental differences may be confounding heritability estimates.

To extend the card metaphor, the similarity would not derive from the luck of the shuffle but because the 23 decks systematically varied in the frequency of the various suits they contained.

The authors showed the results of this exercise in a table in the supplementary material. Some chromosomes were statistically correlated with some others; that result is to be expected by chance. But the authors never calculated the statistic for whether what they observed represented an overall shift toward chromosome relatedness moving in tandem across pairs of individuals. A simple bar chart, reproduced in figure A2.1, shows that indeed there was an overrepresentation of correlated chromosomal pairs. If this were truly a game of chance, then we would see about 10 percent of pairwise comparisons fall under the statistical probability of being present less than 10 percent of the time. Instead, we see a shift to the left, to rarer (more correlated) pairs of chromosomes in their data. A statistical test to see whether this overall pattern could have been obtained by chance showed that it was highly unlikely. Only 1 try in fewer than 10 million would yield a distribution like this by chance. So their key assumption that the differences in genetic similarity resulted only from deck shuffling and not from population-level differences seemed to have less than a one in a million chance of being true.[4]

Sensing a chance to avenge social science's earlier defeat at the hands of the twin analysis, we recalculated the heritability of a range of traits from height (presumably the most unaffected by environmental differences that might confound estimates) to education (which should be among the most affected by environment), accounting for the environmental differences captured by the urban/rural split. Again the heritabilities barely budged. It could be the case that we had the wrong environmental measures, so we tried the "kitchen sink" approach. But the change in estimates was always minimal. The geneticists won again, despite themselves.

APPENDIX 3

ANOTHER ATTEMPT–COMBINING PRINCIPAL COMPONENT ANALYSIS WITH FAMILY-BASED SAMPLES: A POTENTIAL WAY FORWARD (NOT YET ATTAINED)

The coup de grâce to our efforts to test whether the estimated heritabilities for social traits were biased upward came from a study of siblings also made possible by the genomics revolution. As we have mentioned, assortative mating aside, siblings are, on average, 50 percent related. But this is a mere average around which there is considerable variation. In fact, if you have ever felt like you are much more like one of your sisters than another, you may actually be correct. Due to the random shuffling of the decks (the two parental chromosomes) during recombination, when DNA is passed back and forth between the two parental chromosomes to make a single chromosome for the offspring, some siblings end up with more similar DNA than others.[1] As with the GCTA analysis described above, we can then correlate the level of genetic similarity with the level of phenotypic similarity in order to get a heritability estimate.

But as we know, if your parents are genetically similar to each other due to genetic assortative mating, you are going to have more identity by state (IBS) with your sibling than if your parents differ. So merely contrasting sibling pairs across families fails to approximate the ideal experiment because there are two sources of variation in genetic similarity: assortative mating and chance. So we cannot be sure that siblings who are more related did not come from environmental as well as genetic backgrounds that were more similar as well, thus confounding our attempts to get a more defensibly pure measure of genetic effects.

If we had families with three or more siblings, we could perform an analysis that examines the genetic and phenotypic differences between sibling pair A-B and sibling pair B-C within the same pedigree. We could thus be sure that the differences in how alike they were in terms of IBS would be due to chance and not to mating dynamics at the parental level. This method, like all statistical models, requires assumptions, one of which is that the heritability we estimate for larger families also applies to those with just one or no sibling. We also have to assume that the phenotypic variation between pairs of siblings maps onto that between families. The variation in phenotypic distance between sibling pairs from the same family does not have to be equal to that between the randomly selected pairs of non-kin individuals in GCTA analysis or even to the variation across sibling pairs from different families (and it probably would not be equal because the shared family environment would reduce differences in outcomes). As long as the distribution of phenotypic variation in outcomes was as equally truncated as the genotypic variation of within-family sibling pairs (and we assume that the effects of genetic similarity are more or less the same across the entire distribution, i.e., linear), then we can use the more limited within-family sibling-pair differences in both genotype and phenotype to estimate heritability on a larger scale. This approach also assumes that the effect of genetic differences within families is generalizable to such effects between families—that is, that families do not accentuate or mitigate genetic effects in their midst more or less than society writ large does. (This assumption is, in turn, challenged by the notion of niche formation within families and the tyranny of small differences in a household unit. We, in fact, find that effects on educational outcomes for a given genetic difference are actually larger between siblings than they are when comparing individuals from different families. We discuss this more thoroughly in chapter 3.)

Very few data sets have adequate numbers of large families in which three or more siblings are genotyped. We can only think of one such data set, the Framingham Heart Study, and it does not have enough of those families to estimate such a model. (We tried.) But there are other potential approaches that would do as well in mimicking the experimental, random assignment of genotype. One would be to measure both parents' genotypes at each locus, generate a predicted offspring genotype at that locus, and then calculate the deviations from that average, expected genotype. These would, in essence, be random and thus purged of any association with either the environment or assortative mating. In other words, if at a given location both your parents were GG, then you would be predicted to be GG and you would, by definition, be GG. Thus, that particular position would add nothing to our endeavor. But if at another position, both of your parents were GC, then you would be predicted to be GC as well, but if you ended up homozygous (CC), then you would be +1 [C] or –1 [G] depending on how you coded the reference base. We could then take these "residualized" genotypes to calculate sibling IBS across families and see how they predicted phenotypic resemblance. Or, we could perform GCTA analysis with unrelated individuals, since we would have purged the data of environmental or assortative mating influences by factoring out the parental genotypes and relying only on the variation that resulted from the last meiotic division.

This sounds like a good approach, except that data sets with the genotypes of both parents and an offspring (known as trios) are hard to come by, especially when you want them to be nationally representative. Thus, another approach can be used to exploit the randomness of recombination and segregation, and that is to measure identity by descent (IBD) rather than IBS. Identity by descent means not only that two people share the same allele type at a given location (G, C, A, or T), but also that they have copies of the very same base passed down through the generations. So if a mother is GC and the father is AT, and the two children are both CT and CT, we know that they are IBD = 2 at that position because their Cs had to be the same C coming from the mother and their T had to be the only T that the father carried. But if the parents are CC and TT, respectively, and the children are CT and CT (which they would have to be), they are only IBD = 1 since we cannot know for sure which C they got or which T. So, because there is a 50 percent chance that they got the same C (donated by the same grandparent), we give them a .5 for their simi-

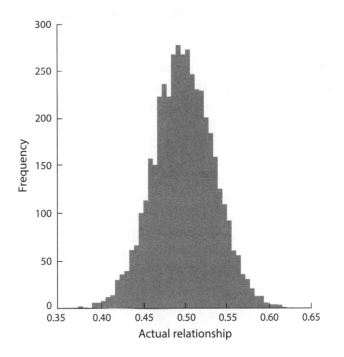

Figure A3.1. The distribution of percent of alleles identical by descent for full siblings in a sample from the Framingham Heart Study.

larity in the indeterminate C and another .5 for the similar scenario with their T. When we do this across the genome, we can calculate the overall percentage of IBD. If we have genotype information on the parents, we can be more certain about our assignment of IBD—that is, that a given shared allele came from the same source. In fact, the more information we have about the entire family tree (pedigree), the easier it is to deduce that alleles are IBD. But when we are not sure, we can still assign probabilities based on allele frequencies in the population. And although riddled with more error (because of being measured less precisely), this approach still generates a randomized distribution that is free from concerns for both assortative mating and population stratification. Indeed, as we can see in figure A3.1, there is significant variation around the "average" 50 percent IBD figure.

Even if you come from the same mother and father, you may be as much as 50 percent more related to one brother as you are to another if one of your siblings is 60 percent IBD with you and the other is 40 percent. Thus, we now have a handy way to measure only the randomly assigned part of genetic similarity and see how well that predicts the phenotypic similarity between sibling pairs. The result is heritability estimates that mimic that of twin-based estimates (~.8) even better than the GCTA approach! The original paper that attempted this approach used Australian fraternal twins to investigate height. (Height is generally seen as a good test case because it is easy to measure and is assumed to have a large genetic component.) We did the same for socioeconomic outcomes with data from a sample of Swedish fraternal twins. (Datasets with siblings from the United States do not have enough cases.) Again we fail to slay the genetic beast: the heritability of education is lower than the twin estimates but higher than GCTA, at 25 percent.

APPENDIX 4

A TURN TO EPIGENETICS AND ITS POTENTIAL
ROLE FOR MISSING HERITABILITY

One other theory of missing heritability has included the notion that epigenetic marks are also heritable and can explain variation in outcomes. Epigenetic marks are chemical attachments to DNA and the proteins (histones) that DNA is coiled around in its stored state (which together form heterochromatin). They are one of several mechanisms to regulate gene expression—that is, when and where RNA is copied from DNA in order to produce the gene product (typically a protein). A methyl group (CH_3) can be attached to DNA wherever a C is followed by a G. Such CpG sites (the "p" stands for the phosphate in the backbone between the bases) are disproportionately found in the regulatory areas of genes: the promoter regions before the start of the coding region (the part that gets copied) and the introns (parts within the coding regions that are excised at the RNA stage).[1] When a CH_3 is attached to the C in a CpG sequence, it tends to reduce the probability that the associated gene will be transcribed (the CH_3 turns down or off the function of the gene). Creating a CpG island—a region with lots of Cs followed by Gs—produces a regulatory hotspot. Methylation is a fairly stable mark and helps prevent deamination of the C (which converts C to U, thus causing a problem in the molecular machinery of the cell). We can think of the CH_3 sticking off the methylated C as a speed bump that makes it more difficult for the DNA transcription machinery to proceed along the strand and do its work. Thus, highly methylated genes tend to be silenced.

This turning on and off of genes at different times and places is critical to development since each cell—from a neuron to an osteo-cyte to a liver cell—all share the same DNA sequence common to the individual. Thus, it is epigenetic marks like methylation and histone acetylation (along with other factors like transcription factor gradi-ents; spatial differences in the concentrations of key, gene-activating proteins) that orchestrate tissue differentiation—turning some cells into fingernails and others into neurons. While methylation tends to increase over the lifespan of an organism (allowing for some re-searchers to construct an epigenetic clock to ascertain the "true" age of tissues or a person as opposed to chronological age), it is also re-sponsive to environmental changes (see, e.g., figure A4.1).[2]

A well-documented phenomenon is that when a pregnant mouse is stressed, her offspring tend to be born with more methylated pro-moter regions of the glucocorticoid receptor gene. Since the gluco-corticoid receptor acts as the off switch in the hypothalamus for the release of cortisol (a major stress hormone), when the pups are born, they tend to have a more elevated stress response since they have fewer off switches expressed. In other words, they are on hyperalert all the time. The mother communicated information biochemically to her offspring about the fact that they were likely to be born into a highly stressful world and thus should maintain a higher state of arousal to survive. Such hyperarousal, however, comes with long-term costs. While methylation is stable, it is not permanent, like a muta-tion, so when these pups are adopted out to a very calm mother who spends lots of time licking and grooming them, the marks are erased, cortisol levels drop, and the baby mice calm down. Methylation is not a short-term response, but rather a medium- to long-term one. We do not necessarily methylate our glucocorticoid receptor from one stressful experience of our boss yelling at us, but repeated exposure to an abusive colleague day in and day out may lead to a change in our epigenome.

This detour into the mechanisms of epigenetics circles back to the missing heritability problem because some thinkers posit that epi-

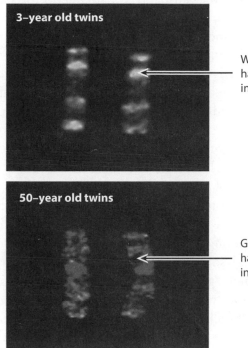

White shows where the twins have epigenetic tags in the same place

Gray shows where the twins have epigenetic tags in different places

Figure A4.1. From M. F. Fraga et al. Epigenetic differences arise during the lifetime of monozygotic twins. *Proceedings of the National Academy of Sciences* 102, no. 30 (2005): 10604–10609. Copyright (2005) National Academy of Sciences, U.S.A. Source: http://learn.genetics.utah.edu/content/epigenetics/twins/. Text box: Chromosome 3 pairs in each set of twins are digitally superimposed. The 50-year-old twins have more epigenetic tags in different places than do 3-year-old twins.

genetic marks can be inherited across generations. If this were the case, it would come as no surprise that much heritability is missing since we do not typically measure epigenetic marks. We usually measure genetic variants, but not whether they are turned off or on. If such large-scale, environmentally induced epigenetic inheritance were the case, it would represent a huge revolution in molecular biology and the complete restoration of Jean-Baptiste Lamarck to the pantheon of biology.

Namely, epigenetic inheritance would revive the idea that environmental influences on development could be transmitted to the next generation. Epigenetic methylation marks are indeed preserved and inherited in mitosis—when one cell splits into two. However, in the process of meiosis, which produces the reproductive cells of sperm and ova, epigenetic marks are erased from DNA. This is critical because when fertilized, the zygote must be an omnipotent stem cell; that is, the new cell must be able to turn into any kind of cell in the body since it is the mother of all sorts of cells, from osteocytes to leukocytes and beyond. So for transgenerational memory of epigenetic marks to be possible, these marks need to be selectively and incompletely erased or the relevant information about their restoration must be transmitted through some mechanism other than the marks themselves—an epi-epigenetic ghost in the machine, so to speak, an entire system of information transmission that is parallel to the inheritance of genetic code. If such a mechanism exists, we have not yet found it, although we may have found bits and pieces of such a Rube Goldberg–like contraption.

One such violation of Mendelian genetics occurs in the process called imprinting (also known as parent-of-origin effects). At least 30 genes in human beings have shown signs of imprinting, which involves the expression of only that copy of the gene inherited from a particular parent. So while you may get two copies of a particular gene, one runs the show in your body. Many imprinted genes are expressed in utero, where there is a battle between the evolutionary interests of the father and the mother. The father, not knowing whether he will ever be able to successfully impregnate the mother again, wants the baby to grow as large and as robust as possible at the expense of the longer-term fertility resources of the mother (since the next baby she could have might not be his). But the mother, who is making the huge metabolic investment in the fetus (hence the old adage, lose a tooth for every baby), wants to hold some reproductive resources in reserve in case this baby with this particular father turns out to be a dud. This battle royale largely plays out in the placenta, which has the genotype of the child, not the mother.[3]

It turns out that genomic imprinting does not stop there, though.[4] Certain areas of the brain overexpress paternal genes while other areas overexpress maternal genes.[5] Your mother and father are literally struggling to control your mind! For imprinting to work, the off-spring brain needs a way to remember which gene (or chromosome) came from which parent, so there must be a mechanism for doing so. Interestingly, there is no evidence so far for grandparental effects: your mom got one copy of a maternally imprinted gene from her mother and one from her father, but when she passed it onto you, it did not seem to matter whether she gave you the one that she never used (her father's) or the well-tested one she got from her mother. So it seems that with imprinting, the transgenerational train stops at only two stations.

As tantalizing as imprinting is, the ability of environmental infor-mation to be written into the epigenome and transmitted across gen-erations is another order of magnitude more complicated. To revisit the glucocorticoid and stress example, the offspring's cells must know to methylate the glucocorticoid receptor because of the environment the relevant parent experienced, even though this is not an imprinted gene. The child's cells must also know to do this in the hypothala-mus, because methylation is cell-type, location, and timing specific. Such memory would require a code that is even more complicated than DNA itself. Transgenerational memory could be very useful in changing environments, but it seems very costly from an evolution-ary perspective. More realistic would be that a few critical genes might be subject to the influences of the parental environment.

At least one study of mice claims to have demonstrated such trans-feral of environmental information across generations as seen through epigenetic profiles. (Just showing correlations between parents and children in their epigenome is not sufficient evidence because their shared genotypes, as well as their shared environments, influence methylation.) In this experiment, male mice were exposed to an odor that was paired with a shock. The result was a specific freezing-up behavior when they were exposed to the odor—just as Pavlov's dogs salivated at the sound of a bell that had been rung when they were about to be fed. This behavioral response was associated with a certain methylation profile on the gene that expressed the receptor for that particular olfactant in the nasal tissues.

The full-treatment mice were then mated both with typical, untreated female mice and with another group of control female mice that were exposed to only the odor and not to the associated shock (and thus did not show the same epigenetic profile). The offspring of the mice seemed to have similar epigenetic profiles for the genes that coded for these receptors in the same tissue location as their fathers' (i.e., their noses). And more critically, the offspring of the full-treatment mice seemed to display the same response to the odor that their fathers had been conditioned to perform (albeit to a weaker extent). Fathers are deployed rather than mothers because mothers have several ways to transmit information, such as uterine conditions or what RNA they deposit in the egg. It is perhaps possible that some paternal RNA enters the zygote through the head or piece of the tail of the sperm, but proportionately by mass it would have to be much, much less, if any.[6]

Such studies are very exciting and whet the appetite of social scientists troubled by the one-way arrow from genes/biology to social behavior implied by the traditional, Darwinian model of inheritance. If what the researchers claim is indeed true, the social environment is again playing a significant role, and we may see the scars of generations past (or at least one generation past) appear in offspring. Social scientists have jumped on the epigenetic bandwagon in droves since they tend to be much more comfortable with a narrative that has social environment as the cause and biology as the effect, as opposed to the reverse.

Indeed, there have been many studies showing correlations of grandparental environmental conditions with grandchildren's outcomes; however, these have not been experiments, so a much more plausible explanation is that the social outcomes of the grandparental generation condition the environment of the parental generation, which in turn, affects the phenotypes of the grand-filial generation. In other words, sociocultural and economic inheritance are much more likely to be at work here. And as for the mouse olfaction study, it has yet to be replicated, so many scholars are skeptical of the results. Yet there are many known pathways for the transmission of information across generations. There is incomplete erasure of epigenetic marks (which is thought to be a mechanism in imprinting) and the reestablishment of them. Mothers also deposit a host of molecules into the ovum, ranging from RNAs in the form of p-bodies (tightly packed for storage purposes) to short RNAs called piRNAs to fatty acids to prions (of mad cow disease fame). These can help guide the epigenetic erasure or reestablishment or, alternatively, affect offspring phenotype directly. And, of course, through nutrition and hormone levels (some of which may be stress related), mothers are in constant communication with their in utero mammalian offspring about the environment that awaits them once they enter the world. Even fathers may contribute information other than their nuclear DNA sequence, since it is now known that portions of the tail (which contain, for instance, paternal mitochondria and some sperm-associated proteins) penetrate the zygote.

The above-mentioned pathways all represent intergenerational effects rather than transgenerational effects. The distinction is between an effect that is directly mediated through active chemical communication between overlapping generations (inter-) and one that lasts across multiple, nonoverlapping generations (trans-) and is not reestablished through communication (chemical or cultural). In the case of mothers, since her daughter's ova are formed while she is still in utero, we would need to go to great-grandchildren to make sure we are seeing a true, transgenerational effect. In the case of fathers, we would need to see it in the grandchildren.

Many studies in plants have found evidence of environmentally induced effects that persist for multiple generations.[7] If we were going to find transgenerational effects anywhere, it should be in the plant kingdom, since plants are typically sessile, and in many species, seed dispersal is over a limited range. This means that any information about the environment that ancestor plants can provide to descendants may be very useful—at least compared with animals who often migrate to new horizons. But even these have been controversial and replication has enjoyed mixed success. A discussion of one of the best human-based studies will help elucidate why.

In Överkalix, Sweden, the descendants of 303 individuals who had been born under differing (and documented) nutritional conditions in 1890, 1905, and 1920, and who themselves engaged in different behavior (such as smoking), were followed until 1995. In typical human studies claiming a transgenerational effect of environmental conditions of grandparents, it is impossible to rule out step-by-step cultural transmission. For example, the grandchildren of those who were in utero during the Dutch Hunger Winter of 1944, when food was scarce due to Nazi occupation, evince higher rates of cardiovascular disease late in their lives. However, this could be due to an effect on the children's dietary preferences—say having more voracious appetites for food high in saturated fats to "make up" for their in utero deprivation—which is then transmitted culturally to their children at the breakfast table. What makes the Överkalix study more convincing is that the transmission was sex-specific in the grandchildren even though it was entirely measured through the paternal line.[8]

In other words, the paternal grandmothers' conditions affected the granddaughters but not their grandsons, while the paternal grandfathers' circumstances affected their grandsons. The idea is that since these effects are going through a single pipeline—the father—they cannot be attributed to differential cultural inheritance. Of course, that relies on a very simple notion of how cultural inheritance works. What if grandparents socialized their grandchildren—as they do in many cultures while the parents work—and they did this in a sex-specific way? Or if the fathers treated their sons and daughters differently based on their own condition? The family is not a simple photocopy machine but rather a Rube Goldberg–like contraption.

Even if we could wave a magic wand and eliminate cultural differences, there would still be other mechanisms that would confound our attempts to isolate an epigenetic transmission pathway. For example, we know that the epigenome is largely determined by the genome. And we know that it is not entirely random which sperm and ova succeed in producing a live baby (let alone one that reaches sexual maturity and actually manages to produce grandchildren). For example, it has long been known that uterine stress increases the rate of spontaneous abortion (miscarriage), and that boys are more sensitive to stress such that after stressful conditions—such as war or natural disasters—we see the sex ratio of live births skew more toward girls.[9] It is also thought that more robust genotypes survive certain conditions over others. So any multigenerational effect could be due to sperm competition and selective survival, both in plants and in animals.

Thus, even if we could experimentally manipulate the environment and control which offspring mated with which, we still could not be sure we were not subtly selecting for genotype! This holds true not only for humans but for lab animals and plants as well. What we really need is a set of twins or clones in order to keep genotypes entirely constant. Back-bred, isogenic lab strains approach this ideal, but certainly in humans—absent a mad scientist who captured a whole slew of twins and manipulated their breeding and environments across generations—this seems unlikely. Thus, it may be a long time before we know whether the environmental traumas of our ancestors are written into our epigenome today. Stay tuned on the issue of transgenerational effects.

In the meantime, even if such Lamarckian inheritance does indeed occur, it probably cannot account for missing heritability. Recall that most heritability estimates come from either twin-based models or GCTA. Both monozygotic and dizygotic twins share the same uterine environment and maternal communication (though dizygotic twins do have different placentas and monozygotic twins may or may not share one). If the mother was stressed, either in utero or earlier (or her mother was), we should see those effects passed to both her fraternal twins and her identical twins, thus it would not explain the greater similarity of monozygotic twins than dizygotic twins, which is the basis for heritability estimates. Likewise, if a transgenerational epigenetic effect were to manifest as heritability in a GCTA model, it would need to persist through many (>8) meiotic divisions, since that is how distantly related the pool of "unrelated individuals" is in such analysis. Even among plants, the most persistent effects show up, at most, for four to five generations.

It is possible for the epigenetic story to inflate heritability estimates in adoption studies, in which heritability is based on the biological parent-child correlation. But here epigenetics is the least of the problems: prenatal conditions are more likely to result in interpreting those associations as genetic in nature. Likewise, it is theoretically possible that the sibling IBD method discussed in appendix 3 also picks up epigenetics, if the epialleles segregate with the IBD chunks of the genome. Of course, we have to keep in mind that since all of these different approaches converge on the same heritability estimates, more or less (with the exception that GCTA tends to be lower), we should probably conclude that epigenetics is not likely to fill in the gap between measured genetic effects and overall additive heritability.

APPENDIX 5

ENVIRONMENTAL INFLUENCES ON RACIAL INEQUALITY IN THE UNITED STATES

There is, of course, no end to measures that capture racial inequality in the contemporary United States. And as we have seen, even defining race in America is no clean and simple task. Even so, we should review some statistics that contrast the status of non-Hispanic African Americans with non-Hispanic whites to illustrate the extent of racial inequality in contemporary America. These are based on self-identification in survey and administrative data sets.[1]

In comparing blacks and whites, we can observe that blacks are about half as likely as whites to hold a bachelor's degree. Probably due in no small part to this education gap, African Americans are twice as likely as whites to be unemployed, and among those who are employed, they hold professional or managerial occupations only half as often. Blacks earn, on average, 70 cents on the dollar of white wages. An even starker result of these inequalities in education and the labor market, African American wealth (net worth) is less than one-tenth that of whites. (This difference is reduced but not eliminated if we compare blacks and whites with the same income level.) In addition to these socioeconomic differences, there are huge disparities in other outcomes like family structure, health, and academic achievement.[2]

There are several plausible explanations for the large racial disparities we observe. The legacy of past racial oppression may linger, either in terms of culture or in terms of economics. Studies have found that when we compare blacks and whites from families with similar parental education and levels of wealth, they perform similarly in school and in the labor market.[3] In that case, it could just be a matter of time—albeit generational time—before gaps narrow completely, as black wealth and education levels catch up to those of whites. The problem is that the gaps are not narrowing. In terms of education, it is a tale of two genders. Black women continue to make some progress vis-à-vis their white counterparts, but black men are stagnating in the education system and labor market. The story for wealth is even more disheartening. The black-white wealth gap is bigger now than it has ever been. So what explains the lack of progress on these fronts?

Conservatives argue that a self-defeating oppositional culture helps account for why U.S. blacks—and other minority groups that did not willingly come to these shores—fare worse than the majority group. The theory is that pursuing mainstream avenues of success is viewed as "acting white." Because such behavior implies an adoption of the cultural practices of the oppressor, blacks and others rebel against academic achievement (and other middle-class, white values), leading to a self-defeating culture of poverty or "getting by." First introduced by the anthropologist John Ogbu, this oppositional culture thesis is elegant in that it explains both voluntary immigrant success and involuntary immigrant failure to thrive socioeconomically.[4] Those who come to this country voluntarily actively embrace the dominant institutions in society—think Asian Americans—but those who have been either conquered (Native Americans or Central Americans) or been brought here in chains (African Americans) go in a defiant direction. This theory accounts for the success of first-generation African or West Indian immigrants (who fit the voluntary model), while also accounting for the downward mobility of subsequent generations as they come to identify and be identified with the African American experience. There are various flavors of such culture explanations, including ones that focus on concentrated poverty (rather than immigration status) as the key factor.

Liberals argue that such theories blame the victim, and we should instead focus on discrimination and racial bias in society. By their account, some of the inequality producing dynamics may have race-

neutral origins but end up having disparate impacts. Take, for example, how people get jobs. For most occupations, hires are not made cold. That is, most individuals who are successful in landing a position did not come into the firm via the want ads. Rather, as the sociologist Mark Granovetter showed in 1973, it is through connections that most of us obtain gainful employment.[5] And not just any connections: it is weak ties—the people we are friendly with but do not necessarily see all that often (acquaintances)—that are most important. They provide us with new information that our close friends and family do not, since we typically already know the same information as our close friends do. From the employer side, it makes sense to hire someone who is recommended by a current employee over a complete stranger for many reasons. First, you have more information about the friend of a friend; you can perform a preinterview of sorts with your employee. Second, you have your friend's reputational vouchsafe: if the new person turns out to be a complete dud, it is on your connection's head, so that person has a strong incentive to recommend only people whom she thinks will really succeed. Finally, we all want to work with people who are culturally similar to us, and the best way to insure that is to hire those who are already in our social networks. While all of these rationales make sense, they lead to institutionalized occupational segregation by—you guessed it—race, which, in turn, reinforces existing inequalities. For this very reason, in some industries—like the public sector, for instance—there exist quite formalized, color-blind approaches to hiring such as the civil service exam. Perhaps, then, it is not surprising that blacks are more likely to work in this sector.

Additionally, it is possible that there is ongoing, purposive racial discrimination in today's society that prevents minorities from realizing their full potential. A number of studies have suggested that contemporary discrimination is a key factor for employment differences by race. Experiments that swap out black and white names on otherwise identical resumes have found that the black names receive far fewer call backs than the white names do.[6] Similar audit and correspondence studies have been conducted for face-to-face job applicants, for home purchasers, and even for auto-mechanic quotes.[7]

Black-sounding names have also been found to disadvantage kids in school—even compared with their own siblings.[8]

Such purposive discrimination could be either overt or statistical. That is, it could be driven by a social psychological dynamic by which employers (or real estate agents, or teachers) have a visceral (perhaps unconscious) reaction to African Americans or other minorities and simply do not want to interact with them. Indeed, studies that try to assess unconscious bias, also known as implicit bias, find that not only whites but also blacks themselves have internalized racial stereotypes that lead respondents to associate blacks with fear, disgust, and other negative emotions.[9] Such unconscious bias is measured by an implicit association test in which respondents are timed for microsecond differences in pressing keys to match concepts, words, faces, and the like. Subjects tend to respond more slowly when black faces are paired with positive attributes and more quickly when black faces are paired with negative attributes—and vice versa for white faces.

It is hard to know to what extent conscious or unconscious irrational feelings may be driving real, consequential behavior in the economy. A devastating dynamic effect of these irrational negative stereotypes, just like oppositional cultural dynamics, is that they can create their own reality. This is most famously illustrated by the classic 1960s study *Pygmalion in the Classroom*,[10] in which researchers told teachers that certain (randomly selected) kids in their classes had tested in the superior range on a powerful new IQ test. By the following year, those "chosen" kids demonstrated real cognitive gains over their unselected peers. There was, of course, no secret, super-duper test. The mere halo effect of teachers (and parents and others) thinking these kids were prodigies led them to be treated (and, in turn, perhaps behave) differently, which led to real advantages. Likewise, when Claude Steele and Joshua Aronson primed black students with negative stereotypes about themselves (by reading them an announcement about racial test score gaps) before giving them an SAT-like assessment, the treatment group (that was negative-stereotype primed) performed significantly worse than the control group of African American test-takers.[11] (Similar effects have been demonstrated for whites vis-à-vis Asians and for women vis-à-vis men with respect to math.) So our irrational beliefs can create their own reality and thus seem logical.

While it has become socially unacceptable for individuals to overtly express racial prejudice, and surveys show that racist attitudes have declined sharply since the 1960s, it is hard for social scientists to know for sure whether such trends represent true "changes of heart" or, rather, are merely the result of racist respondents learning to hide their true views to avoid being shunned.[12] This is what survey researchers call social desirability bias. It is not limited to race, of course, but extends to any charged subject such as gender or even obesity.

There are several ways to try to distinguish between political correctness and true racial harmony. One of the ways is the implicit association test mentioned earlier. Another is election results. And they paint a rosier picture. Yes, we Americans elected a black president, but more important, we elected him by the exact margins predicted by the polls. Political scientists first noticed a large difference between survey-based calculations of political support and actual voting outcomes back in 1982, when Tom Bradley, the black mayor of Los Angeles, was running for governor of California against a white Republican candidate, George Deukmejian.[13] Polls heading into election day had Bradley cruising to victory, but he lost. A similar thing happened in 1993 to the black mayor of New York, David Dinkins, in his race against Rudy Giuliani. The polls were wrong.

These were cases of social-desirability bias striking back. When pollsters queried likely voters, they got one story. But when voters were in the privacy of the voting booth, they voted their true hearts. It is probably not a case of sinister racists telling pollsters one thing to fake them out and then sniggering all the way to the polling station. Rather, it is most likely an unconscious process by which support for the black candidate is "softer" than for the white candidate, so turnout among the minority candidate's constituents is lower when it comes time to make the effort to vote after a long day's work. Or perhaps folks change their minds on the relative weight of various "issues" at the last minute.

What is interesting is that while there has been no surge in minority representation in Congress or state legislatures—despite the presence of a black president—the Bradley effect seems to be ebbing. (Or perhaps, as some new research suggests, it never really existed outside of a few high-profile cases.[14]) Preelection polls tend to be more accurate in those cases when a black candidate runs against a nonblack candidate. Obama's margins of victory were exactly what Nate Silver predicted; ditto for former Massachusetts governor Deval Patrick. Meanwhile, the Democratic (African American) candidate for the U.S. Senate in Tennessee, Harold Ford, lost, but he lost by the exact amount the polls predicted he would. Perhaps more tolerant racial attitudes had a large component of political-correctness bias in them as recently as the 1980s, but respondents seem to be saying what they believe today—at least during election season. Of course, there are still plenty of people who have no compunction about expressing openly prejudicial views—to wit, Donald Trump.

An alternative type of labor-market discrimination without sociopsychological underpinnings—what scholars label statistical discrimination—may also be at play. For example, employers may assume that all things being equal (say, on a resume), African Americans have had lower-quality educations than their white counterparts due to the reality of school segregation and inequality in districts. Better schooling, the manager reasons, means a higher skill level. This sort of rationale is called statistical discrimination because the perpetrator does not have any negative personal feelings toward African Americans but is instead using her knowledge of statistical differences between groups (that African Americans on average attend worse schools than whites) to infer that a typical African American applicant will be less qualified (because of this inferior schooling, rather than because of innate characteristics) than a typical white applicant.[15] Again, this is a form of institutional racism in which existing occupational inequalities are perpetuated by the "system" doing what it does to be efficient (hiring those with the greatest likelihood of being good employees) in a capitalist society. There is no big conspiracy on the part of whites, but no great effort to overcome existing inequalities, either. The worse labor market outcomes, in turn, lead to blacks living in poorer neighborhoods and their children attending lower-quality schools. "So it goes," in the eternal words of Kurt Vonnegut.[16]

APPENDIX 6

IMPUTATION

The Bell Curve was written before the genomics revolution, so answering the questions posed by studying the interplay of race, IQ, and genetics was not even possible back then. But today it just might be. What if we took the polygenic score for education, discussed in chapter 2, and swapped it for IQ in their analysis? If blacks (on average) scored lower than whites on this dimension, it might tell us that there is indeed a genetic basis for racial differences in education and cognitive ability—or would it?

First of all, the score predicts 6 percent of the variance in education or cognitive ability among whites at the present time. So it is a very poor proxy for the 40 percent of IQ that is supposedly genetic. The other 34 percentage points of genetic variance in IQ may follow a distribution that is quite different from the 6 percent that we observe. This possibility is increased by the fact that there are different linkage structures for Europeans than for those of African descent. Recall that the score was calculated from the analyses of hundreds of thousands of individuals across many studies in myriad countries. But all of those people were white. The genetic markers that have been typed by the big genotyping chip companies, like Illumina or Affymetrix, are the ones that have shown common-enough variability in the human population (markers for which there are typically only two versions—such as G or T—at the given location and for which the less frequent allele appears at least 1 percent of the time). Since the most common consumers of genotyping services are Europeans and other associated white people (for example, about 77 percent of 23andme clientele are European, and only 5 percent are African American[1]), the chips are designed to pick up markers that display common genetic variation in the European-descent population. So, since the polygenic scores are created from white samples, they are picking up relatively fewer common variations when applied to African Americans.

But that is only part of the problem. A bigger issue is that the markers that are genotyped—and therefore on which the polygenic score was calculated—were not necessarily the magic SNPs, that is, the single nucleotide changes that have real biological consequences for brain development or behavior through some other mechanism (say, making someone taller and thus more confident). They are more like flags planted along a transcontinental railroad, spaced along the "chromosomal" tracks to mark given swathes of territory. There are 3 billion nucleotides in the human genome, 1 percent of which show known variation in human populations today. That means there are 30 million sources of SNP-based genetic difference. Of course, most of these are meaningless. The typical chip that is used for genotyping will directly measure 1 million markers (up from a half million a decade or so ago). That means we are measuring just one-thirtieth of the genetic variation in humans.

Thanks to the 1000 Genomes Project (as well as to the earlier Hap-Map Project), we are not limited to the 1 million measured alleles. Using a sample of humans from various parts of the world, 1000 Genomes performed next-generation, full-genome sequencing of 2,500 individuals (the name "1000 Genomes" came from the first phase, which included 1,092 individuals). Next-generation sequencing involves multiplying an individual's DNA and then fracturing it into many random, tiny pieces and reading those pieces. To achieve a full "manuscript" of an individual's genome (all 3 billion base pairs) requires about 28 reads, or passes. But to detect most variation (present in at least 1 percent of the population) four reads is generally enough. Four reads per person times 2,500 people means that the project is able to pick up variants that occur in at least 1 percent of the population. For coding regions of the gene, they performed additional reads to get more "depth" (i.e., rarer variants). And unlike the commercial chips that take a limited sample of such variation, 1000 Genomes pretty much gets it all. And to boot, it gets it all not just for white folks but also for populations ranging from Han Chinese to the Mende in Sierre Leone to the Kinh in Vietnam.

When the various studies conduct their analyses to generate their contribution to the combined score through the process of meta-analysis, they impute their measured SNPs to the 1000 Genomes platform so that everyone—regardless of which chip they are using—has more complete coverage of the same alleles. (Theoretically, at least, because some studies may have missing data or other variations that prevent some alleles from being imputed.) Imputation involves using the flags that have been planted along the railroad tracks to infer the variants around them through their matching up with specific haplotypes (groups of SNPs that are fellow travelers). So imagine at position 10 of chromosome 1 (the tenth base pair from one end of the strand), the Illumina chip measures for variation between C and A. And at position 1000, the same chip measures for variants T and A. The imputation platform—the sequencing of the positions in between 10 and 1000—will reveal perhaps a half-dozen other markers that show significant variation.

So, for this stretch of DNA, we have four possibilities for each chromosome 1 in our sample, represented by position 10 and position 1000: C and T, C and A, A and T, and A and A. If we find that C and T bracket a specific interim sequence that is almost always in the relevant 1000 Genomes sample, say, ATGGA, then we can impute those variants in between and increase the number of alleles we can add to our score computation. This is an oversimplification, since imputation is not just based on two bases bracketing the middle section, and it does not work as cleanly as always capturing a set sequence, but the basic idea is the same. One benefit of imputation is that different chips can all obtain information on more or less the same positions; another benefit is that we get more information. We get less than if we actually measured that interim region, but we get some marginal increase. (In fact, the added predictive power of using imputed bases in addition to measured bases in the replication sample—the one you want to predict the result for—is very small within European populations.)

Why this huge detour into the land of imputation? The answer is that the imputation is performed within ethnic groups. That is, the score—and most scores for other outcomes as well—have been calculated on data imputed to the European HapMap or 1000 Genomes samples. Recall that the haplotype structure for African populations is much different than for non-African populations. Specifically, there is much more variation for sub-Saharan populations. That means even if the measured genotypes happen to show the same bi-allelic variation in both the European and African American populations (i.e., a C and T, as those anchor SNPs at positions 10 and 1000, respectively), the variation they capture in the middle is likely to be very different between the two groups. Namely, the African populations are going to have many more and more-varied haplotypes—so imputation is going to be more difficult and less accurate for black Americans.

Figure A6.1 shows how this works in practice. One of us logged into our raw data from 23andme and picked a random SNP—this happened to be one on chromosome 8—with the label rs1380994. When we plotted this SNP within the linkage structure of its region for chromosome 8 using the European panel from 1000 Genomes (actually a sample of people of northern and western European descent from Utah in the United States—CEU is the population code), we can see that at a given threshold of linkage (i.e., sorting together), in this case $r^2 = .3$, the window onto which this flag gives us insight touches on four genes plus the intergenic region. This means that if rs1380994 is in our polygenic score, it could be picking up genetic effects that span four different protein-coding regions. Meanwhile, running this same exercise for the same SNP (rs1380994) on the 1000 Genome sample from the Yoruba people in Ibadan, Nigeria, we find that our flag stands for only one protein-coding gene.[2] We get less bang for our buck because of the greater variation in the African sample. And this is for only one tribe from one city in Africa! Meanwhile the American sample is presumably a mélange of ethnicities spanning a good portion of what we call Western Europe. Imagine if

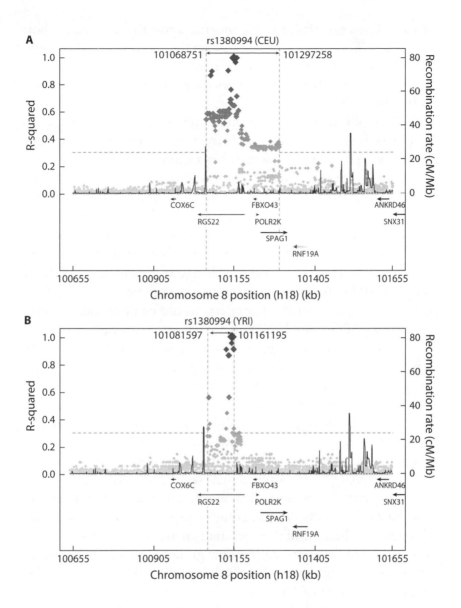

Figure A6.1. Linkage structure for a randomly selected SNP for a European population and a Nigerian population. (A): Utah sample; (B): Ibadan sample.

we had taken a sample from respondents across the same geographical swath of West Africa. The window on the genome that this single SNP would provide would be even narrower still. The point is that the very meaning that a SNP carries differs by race and thus is not really comparable.

This lack of imputable precision matters whether or not we actually use the imputed data or merely the measured alleles. Those flags planted along the railroad tracks may occasionally be placed correctly —by chance—in the spot that matters for the outcome of interest. Perhaps one of the SNPs that Affymetrix measures happens to correspond to position 1 or 2 of three in a codon and thus causes a change from one amino acid to another in the middle of a protein. That protein is a key receptor in the hippocampal region of the brain (among other places in the body), and the change of amino acid affects its ability to bind to the neurotransmitter it is designed to be triggered by in postsynaptic neurons. The effects would be dramatic. The vast, vast majority of measured SNPs do not fall into this category. Instead, they are near enough to some genetic variation that does matter somehow. (And such meaningful differences are mostly in the regulatory machinery of the genome that turns on and off gene expression at various points and in various tissues with more or less efficacy.)

Given the increased genetic variability among African populations, those flags planted along the chromosomes are going to be a lot less informative than they are for the white population (or any non–sub-Saharan population). They are going to tag the relevant "causal" SNPs that matter with less accuracy. This is true whether we impute or not. (And imputation to the African groups represented in 1000 Genomes is less effective because of their greater genetic variation and because the coverage of the continent's ethnic groups is also not as comprehensive as one would like.) The result is that—even if we used chips designed for groups of African descent, ran our discovery analysis on African populations, and imputed alleles using the African reference panel—the predictive power is likely to be worse, absent a much larger sample and more comprehensive genotyping. Even if this is not wholly true (there are, indeed, certain caveats), it remains the case that if we conduct our discovery analysis on individuals of European descent using chips designed for them, then the predictive power of that score in a sample of African Americans is going to be weak. Indeed, those who have tried to deploy the Social Science Genetics Association Consortium measure for education (or even polygenic scores for outcomes like height), have found that they simply do not predict as well for blacks.[3] It is like using a scale to measure length or a ruler to infer weight. Each ancestral population needs a different tool, even admixed ones.

Together, all these issues mean that we cannot merely ask what the distribution of the education polygenic score is for whites and for blacks and then conclude that there is a genetic basis to test score gaps. That would be like weighing one group of kids and finding that they are lighter than another group of kids and thus concluding that the first group of kids are shorter; it may or may not be the case that their lower masses arise from shorter stature. Maybe someday, when everyone in these nationally representative samples has a fully sequenced genome, we can construct scores based on every allele and achieve comparability. For now, that is not possible. But even if it were, that still would not answer the fundamental issue we seek to comprehend.

Eventually we will have large, nationally representative samples that are deeply sequenced so that we can generate polygenic scores that predict IQ or schooling outcomes to a degree that approaches their total additive genetic basis (~40 percent). We may even find differences in the common elements of the polygenic scores for different racial groups. But even then, we will not be able to know the mechanisms—either internal or external—that lay behind their predictive power. That is the inherent trade-off. With a single, candidate gene approach we can attempt to investigate the biological and social pathways that flow from variation. (We may still not be able to map out all the pathways, but we might possibly nail down the major ones.) But the candidate gene approach gives us precious little power to explain the phenotypic variation we observe. There is no way that a single letter change in your genome substantially changes your likelihood of going to college. In contrast, the polygenic score method, by summing variation across the genome, sacrifices any hope of understanding pathways. Even if we find that genes at the top of the list are generally expressed in the brain, we do not know if that is the reason—and not some other mechanism through some other part of the body—they may be associated with IQ. Again, pleiotropy is the rule rather than the exception. Even if they are expressed in the brain and only the brain, we cannot know whether they are associated with tone of speech or a way of walking that itself would have no effect on cognition, save for the fact that it has become racialized and thus evinces differential response from the system that we call society.

Where does that leave us? The empirical prediction one might make from the solid observation that there is more genetic variation among the African American population than among whites is that—all else being equal (which it obviously is not)—we should observe greater variation in continuous, highly polygenic traits such as height or IQ among blacks than among whites. It would be very difficult to draw any conclusions about the mean (average) differences over which all the fuss in the media is made.

Where we do see some real, measurable consequences of these first-order observations about race and genetic variability is in the area of health. One very serious consequence of greater genetic diversity is that it is more difficult to find organ matches for black Americans than it is for white Americans as demonstrated by sociologist Jonathan Daw.[4] Whether strangers or siblings, the greater degree of genetic variability within the black population means that even a brother or sister is less likely to be a viable match than the same pairing for a white donor-recipient dyad. Thus, facile explanations like hospital discrimination or a lack of black familial donors (due to less integrated family structures) probably do not account for the entire race gap in waiting time for a kidney. By looking under our skin and acknowledging the reality of different genetic histories, molecular analysis—when performed without a political ax to grind—can add to our understanding of social dynamics—even in the most fraught and historically dubious area of all, race.

NOTES

CHAPTER 1: MOLECULAR ME: WELCOME TO THE COMING SOCIAL GENOMICS REVOLUTION

1. Ethnocentrists of the nineteenth century sought to use Darwin's concepts of adaptation and natural selection to argue for fundamental biological differences between the groups we call races. When Darwin countered these claims, he curiously found himself allied with his one-time foes: the religious conservatives of the day who also argued for the unity of humankind.

2. We might consider, for example, the last couple of decades of genetics research and social change in the area of sexual orientation. Many queer activists have cheered the search for the "gay" gene in the hopes that if the innate, genetic basis of homosexuality were found, it would increase tolerance since LGBT individuals would be shown not to be making a lifestyle (i.e., a moral) choice about their orientation—they were born that way. In the short term, LGBT individuals may face less discrimination and more rights (like same-sex marriage), but in the long term, earnings inequalities between LGBT individuals and non-LGBT individuals may be further reinforced as "natural" and thus outside the purview of policy.

   Although there may, in fact, exist gay genes, the current consensus suggests a significant environmental component to sexual orientation, further complicating the discussion (R. C. Pillard and J. M. Bailey, "Human sexual orientation has a heritable component," *Human Biology* [1998]: 347–365). There is also the related curiosity from an evolutionary perspective of how a "gay gene" could survive in a population. One idea is that these "gay genes" are actually better thought of as "male loving" genes that are beneficial (fitness-wise) in the population. It has been shown that female relatives (who are likely to have the "male loving" gene) of gay men have more children than we would otherwise expect. See, for example, A. Camperio-Ciani, F. Corna, and C. Capiluppi, "Evidence for maternally inherited factors favouring male homosexuality and promoting female fecundity," *Proceedings of the Royal Society of London B: Biological Sciences* 271, no. 1554 (2004): 2217–2221.

3. R. Herrnstein and C. Murray, *The Bell Curve: Intelligence and Class Structure in American Life* (New York: Free Press, 1994).

4. Furthermore, if our human outcomes and traits are now increasingly explained by discoveries in genetics and some of our basic social constructs become complicated by findings from the biological sciences, will the social sciences lose their relevance in understanding human social behavior? Put another way, if genetic influences can account for a large portion of variation in social outcomes, or if genetics is pointing the way toward new and

better kinds of "racial" and social classification, will we still need sociologists, economists, and political scientists? Meanwhile, genetic data is also pouring into what were once purely social scientific studies, raising old debates about genes and IQ, racial differences, risk profiling, criminal justice, political polarization, and privacy. Discussions and implications about the potential naturalization of these social processes are ongoing, but, like the examples above, are deeply important.

5. An early example was the identification of a gene for age-related macular degeneration on a region of chromosome 1 in 2005. This finding helped targeted treatment for this widespread eye disease that causes blindness in millions of people (http://medicine.yale .edu/news/article.aspx?id=3486); R. J. Klein, C. Zeiss, E. Y. Chew, J.-Y. Tsai, R. S. Sackler, C. Haynes, A. K. Henning, et al., "Complement factor H polymorphism in age-related macular degeneration," *Science* 308, no. 5720 (2005): 385–389.

6. P. McGuffin, A. E. Farmer, I. I. Gottesman, R. M. Murray, and A. M. Reveley, "Twin concordance for operationally defined schizophrenia: Confirmation of familiality and heritability," *Archives of General Psychiatry* 41, no. 6 (1984): 541–545.

7. S. M. Purcell, N. R. Wray, J. L. Stone, P. M. Visscher, M. C. O'Donovan, P. F. Sullivan, P. Sklar, et al., "Common polygenic variation contributes to risk of schizophrenia and bipolar disorder," *Nature* 460, no. 7256 (2009): 748–752.

8. Q. Ashraf and O. Galor, "The 'Out of Africa' hypothesis, human genetic diversity, and comparative economic development," *American Economic Review* 103, no. 1 (2013): 1–46.

9. C. J. Cook, "The role of lactase persistence in precolonial development," *Journal of Economic Growth* 19, no. 4 (2014): 369–406.

10. L. M. Silver, "Reprogenetics: Third millennium speculation," *EMBO Reports* 1, no. 5 (2000): 375–378.

CHAPTER 2: THE DURABILITY OF HERITABILITY: GENES AND INEQUALITY

1. M. Diamond, "Sex, gender, and identity over the years: A changing perspective," *Child and Adolescent Psychiatric Clinics of North America* 13 (2004): 591–607.

2. A. R. Jensen, "Heritability of IQ," *Science* 194, no. 4260 (1976): 6.

3. P. Taubman, "The determinants of earnings: Genetics, family, and other environments: A study of white male twins," *American Economic Review* 66, no. 5 (1976): 858–870.

4. A. S. Goldberger, "Heritability," *Economica* 46, no. 184 (1979): 327–347.

5. Ibid.

6. Recall that Don Draper marries Megan instead of Dr. Faye Miller, although they have no children. Roger and Jane are another example.

7. One could argue that the era of mass incarceration, which we may or may not be exiting about now, is the perfect embodiment of such a policy impulse.

8. It should be said that since the publication of *What Money Can't Buy*, there has been much research using better approaches than observational data that shows that income matters—but not always in the way one might think. Some studies have involved natural experiments, such as the windfall that certain tribes have received thanks to gaming licenses or policy changes that resulted in income "notches" in the social security system. Additional, ongoing research is pursuing the gold standard of evidence: a randomized controlled trial. A group of scholars led by Greg Duncan have obtained funding to liter-

ally give a significant monthly stipend to some poor families (and not others) and track, using the latest in brain scanning and other technologies, how their children fare; however, it will be a long time before the results come in.

9. Indeed, there is social science evidence suggesting that speech patterns are related to earnings in the labor market. Using vocal samples recorded from telephone-based surveys from decades ago, Jeff Grogger has shown that black males who do not "sound black" in their speech earn about the same as white males ("Speech patterns and racial wage inequality," *Journal of Human Resources* 46, no. 1 [2011]: 1–25).

10. https://www.arts.gov/news/2016/arts-and-cultural-production-contributed-7042-billion-us-economy-2013.

11. The "doubling the difference" shortcut comes from the assumptions of the models and the 50 percent resemblance of fraternal twins. The model starts with the assumption that the variation in the trait can be split into the additive elements A, C, and E. Next, we can see that the extent to which identical twins (mz) are similar (the correlation, r, in their traits) is related to A and C (that is $r_{mz} = A + C$), and the extent to which fraternal (dz) twins are similar is related to ½A and C, with the ½ coming from their 50 percent shared genotypes. This means that $r_{dz} = ½A + C$. If we use an algebra trick, we can combine the two equations and generate a simple relationship that $A = 2(r_{mz} - r_{dz})$, which is where we get the "doubling the difference" rule. With survey data, we can estimate r_{mz} and r_{dz}, making it easy to calculate A.

12. In note 11, we already have calculated A and r_{mz}. Since we know that $r_{mz} = A + C$, this means that $C = r_{mz} - A$. So, C is also easy to calculate.

13. What "counts" as the shared and non-shared environment can be unclear to people not familiar with the ACE model. While the house the twins live in would seem "shared" (and it is), to the extent that living in a small/large house affects the twins differently, then this differential effect of the house is part of the non-shared environment. The extent of error in the measurement of the trait (e.g., twins lying about levels of schooling) is absorbed into the non-shared environment, unless twins in certain households are more likely to lie about their schooling, which could then be called "shared environment."

14. R. Acuna-Hidalgo, T. Bo, M. P. Kwint, M. van de Vorst, M. Pinelli, J. A. Veltman, A. Hoischen, L.E.L.M. Vissers, and C. Gilissen, "Post-zygotic point mutations are an underrecognized source of de novo genomic variation," *American Journal of Human Genetics* 97, no. 1 (2015): 67–74.

15. With the caveat that an X chromosome is larger than a Y chromosome in terms of its number of bases and genes and that the mitochondrial DNA always comes from the mother. That said, some research suggests that some mitochondria actually come from the father due to portions of the spermatozoon tail (which contains mitochondria) breaking off in the zygote (F. Ankel-Simons and J. M. Cummins, "Misconceptions about mitochondria and mammalian fertilization: Implications for theories on human evolution," *Proceedings of the National Academy of Sciences* 93, no. 24 [1996]: 13859–13863). Though other, more recent research that deploys full genome sequencing suggests that any paternal mitochondrial DNA may be policed out of the cell (A. Pyle, G. Hudson, I. J. Wilson, J. Coxhead, T. Smertenko, M. Herbert, M. Santibanez-Koref, and P. F. Chinnery, "Extreme-depth re-sequencing of mitochondrial DNA finds no evidence of paternal transmission in humans," *PLoS Genetics* 11, no. 5 [2015]: e1005040). For a review and discussion, see V. Carelli, "Keeping in shape the dogma of mitochondrial DNA maternal inheritance, *PLoS Genetics* 11, no. 5 (2015): e1005179.

16. The alternate world in which there is a smooth distribution of relatedness could exist if there was a lot of inbreeding—that is, incest. Although some societies, such as Pakistan or

the United Arab Emirates, have high degrees of cousin marriage (more than half of marriages are between first cousins), the incest taboo against first-order relatives mating is thought to be universal.

17. One way of justifying the lack of external validity is to appeal to the possibility of nonlinearities of genetic effects.

18. I. Simonson and A. Sela, "On the heritability of consumer decision making: An exploratory approach for studying genetic effects on judgment and choice," *Journal of Consumer Research* 37, no. 6 (2011): 951–966.

19. Ibid.

20. A. R. Branigan, K. J. McCallum, and J. Freese, "Variation in the heritability of educational attainment: An international meta-analysis," *Social Forces* 92, no. 1 (2013): 109–140.

21. J. M. Vink, G. Willemsen, and D. I. Boomsma, "Heritability of smoking initiation and nicotine dependence," *Behavior Genetics* 35, no. 4 (2005): 397–406.

22. At least one earlier attempt used blood typing, but given the small number of categories, it is more likely to lead to false positives for identical twins. L. Carter-Saltzman and S. Scarr, "MZ or DZ? Only your blood grouping laboratory knows for sure," *Behavior Genetics* 7, no. 4 (1977): 273–280.

23. D. Conley, E. Rauscher, C. Dawes P. K. Magnusson, and L. M. Siegal, "Heritability and the equal environments assumption: Evidence from multiple samples of misclassified twins, *Behavior Genetics* 43, no. 5 (2013): 415–426.

24. If among our population everyone has AA at location 6, then that particular place on the chromosome contributes nothing to our calculations because there is no variation. But if another location is pretty evenly distributed, say, 50 percent CT, 25 percent CC, and 25 percent TT, then that locus will contribute significantly to the variation in our genetic similarity index.

25. We discard any outliers who may be cryptically related, that is, actually related in the lay sense of the word, perhaps unbeknown to themselves. Some people in a sample do not know that they are relatives—say second cousins—of one another. Even if introduced to each other, they would not be able to report that they are second cousins; thus, they are "cryptically" related.

26. The missing half may be attributed to various factors, the most important being that the markers we are using to determine how related the individuals are to each other are not the ones that matter for the outcomes necessarily but are just themselves proxies for the "causal" loci.

27. D. H. Hamer, "Beware the chopsticks gene," *Molecular Psychiatry* 5, no. 1 (2000): 11–13, http://www.nature.com/mp/journal/v5/n1/full/4000662a.html.

28. There are new ways that quantitative geneticists detect and adjust for population stratification. One of the most common is to identify the "structure" in the population from the markers themselves. This involves a statistical procedure in which "factors" or scales are created upon which certain alleles load onto while others tend not to. That is, the researcher specifies a certain number of these factors (called principal components, PCs)—say 10—and the algorithm extracts the factor (i.e., the scale with various values for each allele) that explains the greatest amount of variation in the genotypes in the data. Once that is created, that factor or principal component is held constant while the algorithm goes to work on the residual variation in the data. It does this as many times as the number of factors the researcher specifies. These factors tend to reflect latent kinship in the data. As we will show in chapter 5, which is on race, some factors are readily identifiable. Two factors, for example, can generally pull apart clusters of the major self-identified races in the United States. But like a fractal, it seems that nonrandom mating patterns—that is, population structure—emerge within any subpopulation. Even in a nation most would

consider on the homogeneous side—the Netherlands—two principal components readily separate the population into northern and southern (component 1) and east versus west (component 2). (See A. Abdellaoui, J. J. Hottenga, P. de Knijff, M. G. Nivard, X. Xiao, P. Scheet, A. Brooks, et al., "Population structure, migration, and diversifying selection in the Netherlands," *European Journal of Human Genetics* 21, no. 11 [2013]: 1277–1285.)

It is often thought that by factoring out four or more PCs, the remaining variation can be thought of as randomly assigned at conception within that population, thanks to the process of recombination and segregation (like deck shuffling and cutting) during meiosis. Thus, it is assumed that any effect of any allele that is detected would not suffer from the chopstick conundrum but would be truly causal. Of course, the PCs may be sweeping up actual causal effects of alleles that also happen to vary by geography, ethnicity, or some other factor within the data. But better to be on the safe side and make sure that any segregated mating in the past or present (which produces population structure) is factored out since it may actually reflect environmental or cultural differences à la chopstick usage.

We discuss a (second) failed attempt to use PC analysis to reduce traditional measures of heritability in appendix 2. For the truly interested (or masochistic) among you, we describe a third type of investigation (that also has heretofore failed to slay the heritability beast) in appendix 3.

29. D. Conley, M. L. Siegal, B. W. Domingue, K. M. Harris, M. B. McQueen, and J. D. Boardman, "Testing the key assumption of heritability estimates based on genome-wide genetic relatedness," *Journal of Human Genetics* 59, no. 6 (2014): 342–345.

30. We have not even discussed the adoption-study literature, which also confirms, in broad terms, the twin-based heritability figures. Here the Scandinavians and the Koreans have been a godsend. The Scandinavians have maintained meticulous databases, called registries, for most of the post–World War II period. Every birth, graduation, military enlistment, marriage, incarceration, hospitalization, and death is linked to a national ID number. Not only are citizens taken care of from cradle to grave in the social democracies of northern Europe, they are tracked from womb to tomb as well. And for the most part, they accept this. The result is that scientists can observe children who are given up for adoption at birth and link them to their biological parents as well as to their adoptive parents. It turns out that pre-birth factors (genetics plus prenatal conditions, which cannot be separated with this methodology) account for roughly the same amount of variation for outcomes as indicated by the twin models. A lingering concern is whether the assignment of adoptees is not entirely random. That is, if adoption agencies consciously or unconsciously showed a tendency to match light-haired children with light-haired parents, sickly kids with sickly-looking parents, or in any other subtle way caused adoptive parents to be more genetically similar to their adopted children than two randomly selected individuals in the population, that process would confound the whole premise of adoption studies.

The most reassuring evidence on this point comes from Korean adoption agencies. Here, researchers took advantage of the fact that at least some Korean adoption agencies assign babies randomly—that is, based on a waiting list that does not take into account parental characteristics. Because they are randomly assigned, the worries about babies being matched to adoptive families that may more (or less) resemble their biological parents is removed. And scholars find that the estimates of the influence of genes and environment for the Korean adoptees more or less match the effects of environment for the non-Korean adoptees. (B. Sacerdote, "How large are the effects from changes in family environment? A study of Korean American adoptees," *Quarterly Journal of Economics* 122, no. 1 [2007]: 119–157.) He can estimate only the effect of the postnatal environment since he does not observe the Korean biological parents to see how well they correlate with their adopted-out offspring. However, the size of the post-birth environmental effect matches that of the Swedish and Norwegian adoptees, suggesting that the pre-birth component is probably similar as well.

While there are some of the same concerns about external validity for adoption studies as there were with twin studies—that is, whether we can generalize from the special case of adoptees—the fact that this very different approach yields very similar results is either reassuring or disconcerting, depending on which side of the nature-nurture debate you are on. Looking at adopted children's unadopted siblings (both in their biological families that gave them up and in their adoptive families) provides some reassurance on this point because they show similar patterns of resemblance to those without adopted siblings, suggesting that these are not "odd" families. Again, we social scientists must concede that we cannot ignore genetics and biology when seeking to explain human differences in modern social and economic outcomes. The question is, what are the implications of accepting the reality of heritability?

31. This notion is known as the Barker hypothesis, after its founding author, who actually calls it the thrifty phenotype theory. See, for example, C. N. Hales and D.J.P. Barker, "Type 2 (non-insulin-dependent) diabetes mellitus: The thrifty phenotype hypothesis," *Diabetologia* 35, no. 7 (1992): 595–601.

32. A. C. Heath, K. Berg, L. J. Eaves, M. H. Solaas, L. A. Corey, J. Sundet, P. Magnus, and W. E. Nance, "Education policy and the heritability of educational attainment," *Nature* 314 (1985): 734–736.

33. G. Guo and E. Stearns, "The social influences on the realization of genetic potential for intellectual development," *Social Forces* 80, no. 3 (2002): 881–910.

34. Or educational attainment or income for that matter; J. R. Behrman, M. R. Rosenzweig, and P. Taubman, "Endowments and the allocation of schooling in the family and in the marriage market: The twins experiment," *Journal of Political Economy* (1994): 1131–1174.

35. Paul Taubman's response ("On heritability," *Economica* 48, no. 192 [1981]: 417–420). Goldberger's 1979 "Heritability" paper (see note 4) makes a similar point by suggesting that knowing the degree to which variation in an economically relevant outcome is due to genetic inheritance will at least inform us about the extent to which goals of equity and efficiency are in conflict. In making this argument, Taubman echoes Arthur Okun's suggestion (*Further Thoughts on Equality and Efficiency*, Brookings General Series Reprint 325 [Washington, DC, Brookings Institution, 1977]) that in some instances, efficiency may be enhanced by equity-inducing redistribution when equality of opportunity is highly unequal. Of course, what skills (heritable or not) lead to greater efficiency (i.e., higher productivity) in a given moment of economic history is perhaps itself endogenous to larger evolutionary forces and environmental events.

36. That said, the claim that environmental conditions are responsible for a lower heritability of IQ among African Americans rests on the assumption that the underlying genetic variance is the same across racial groups. In fact, we know that due to the population bottleneck during the human migration out of Africa 60,000 years ago, there is much more overall genetic variation within African populations than in other geographic groups. This difference, added to the fact that most African Americans share European ancestry as well, suggests that lower variance in the underlying distribution of genes does not explain the lower heritability estimates for this group. However, it could still be the case that some selection gradient (such as the Middle Passage of the slave trade) led to a narrower distribution among the particular genes that matter for IQ.

Or it could be that there is a higher amount of assortative mating on the unmeasured genes that predict IQ among blacks, which would also lead to lower heritability estimates (by inflating the degree to which siblings share their genome). Finally, there could be a different degree of genetic-environmental covariance within the families of the two populations. That is, if for blacks environments tend to be more equal across degrees of kinship while for whites more genetically related kin tend to share their environments to a greater degree than their black counterparts, then it is not really that the heritability of a trait for blacks is lower but rather that the heritability for whites is overestimated. This latter possibility brings us full circle back to Guo and Stearns' ("The social influences") original hypothesis (albeit a variation on it that they did not consider): muted gene-environmental covariance between, say, identical and fraternal twins may be one mechanism by which the environment suppresses the realization of genetic differences. It is a specific form, however, with a particular policy implication that differs from the possibility that overall common environmental influences are suppressing heritability.

CHAPTER 3: IF HERITABILITY IS SO HIGH, WHY CAN'T WE FIND IT?

1. A primer binds to a specific gene DNA sequence and allows that string of bases to be amplified—copied thousands of times over in a process called the polymerase chain reaction (PCR) in order to be measured.
2. The gene whose regulatory region to which the GFP has been fused.
3. "Why mouse matters," National Human Genome Research Institute, National Institutes of Health, Washington, DC, https://www.genome.gov/10001345.
4. A. J. Keeney and S. Hogg, "Behavioural consequences of repeated social defeat in the mouse: Preliminary evaluation of a potential animal model of depression," *Behavioural Pharmacology* 10, no. 8 (1999): 753–764.
5. Scientists test murine tenacity using a forced swim test: V. Castagné, P. Moser, S. Roux, and R. D. Porsolt, "Rodent models of depression: Forced swim and tail suspension behavioral despair tests in rats and mice," *Current Protocols in Neuroscience* 55, no. 8 (2011): 11–18. Or see D. A. Slattery and J. F. Cryan, "Using the rat forced swim test to assess antidepressant-like activity in rodents," *Nature Protocols* 7, no. 6 (2012): 1009–1014. For effects in humans see A. L. Duckworth and M.E.P. Seligman, "Self-discipline outdoes IQ in predicting academic performance of adolescents," *Psychological Science* 16, no. 12 (2005): 939–944.
6. *MAO-A* recycles neurotransmitters for reuse once they are brought back to the presynaptic neuron. Monoamine oxidase inhibitors were early antidepressants.
7. "Rs909525," SNPedia, http://www.snpedia.com/index.php/Rs909525.
8. D. Conley and E. Rauscher, "Genetic interactions with prenatal social environment effects on academic and behavioral outcomes," *Journal of Health and Social Behavior* 54, no. 1 (2013): 109–127; J. M. Fletcher and S. F. Lehrer, "Genetic lotteries within families," *Journal of Health Economics* 30, no. 4 (2011): 647–659.
9. Keep in mind that researchers had good theories from animal models guiding the locations in the genome they examined, but they did not have good theories about how that would translate to finding an effect on one particular human outcome but not another.
10. There are a number of efforts now to combat publication bias. One is the availability of non-peer-reviewed online archives, such as bioRχiv archive (http://biorxiv.org/about -biorxiv), that allow researchers to post null results. Another is a movement to require researchers to preregister on a website their hypotheses and the tests they intend to conduct

before they perform the analysis (usually when they are applying for funding) so that scholars have to carefully think about the work they are doing and cannot go on "fishing" expeditions. (For a good summary of the debate around preregistration, see S. Mathôt, "The pros and cons of pre-registration in fundamental research," Cognitive Sciences and More, http://www.cogsci.nl/blog/miscellaneous/215-the-pros-and-cons-of-pre-registration -in-fundamental-research.) Finally, there exist efforts to make replication—key to confirming the "truth" of findings—a more respected endeavor ("Estimating the reproducibility of psychological science," Reproducibility Project: Psychology, Open Science Framework, https://osf.io/ezcuj/wiki/home/).

11. More generally, there are issues with preregistration of study designs and statistical tests. A fear is that a requirement of preregistration of all analyses might limit innovative study designs and/or novel hypotheses as well as inductive learning. L. C. Coffman and M. Niederle, "Pre-analysis plans have limited upside, especially where replications are feasible," *Journal of Economic Perspectives* 29, no. 3 (2015): 81–97, http://pubs.aeaweb.org/doi/pdf plus/10.1257/jep.29.3.81.

12. C. F. Chabris, B. M. Hebert, D. J. Benjamin, J. Beauchamp, D. Cesarini, M. van der Loos, M. Johannesson, et al., "Most reported genetic associations with general intelligence are probably false positives," *Psychological Science* 23, no. 11 (2012): 1314–1323.

13. Think of the appealing notion that a gene for aggression lands you in jail if you are from the ghetto but in the boardroom if you are to the manor born.

14. The nice thing about the plot is that not only does it show the most statistically powerful SNPs, it shows them in order. So if a result is a false positive, it is likely to be a lone dot, hovering way above all its neighbors, because it was likely found in error due to chance or a genotyping mistake. However, if it is a "true" result, it is likely to have a bunch of dots in the same area that look like they are climbing up to the peak. These dots are results from independent tests of association for different SNPs with the outcome. Since those that are near the most significant one are also sending strong signals of association, it suggests that that region is truly associated with variation in the outcome. This is due to the phenomenon of linkage disequilibrium (see appendix 6 for a more thorough explanation), whereby SNPs near each other on a chromosome can act as proxies for each other. So the signal gets stronger as we approach the biggest hit and fades as we move away from it. This can be seen clearly in the Jackson Pollock-like drips of powerful SNPs in chromosome 19 at the right end of figure 3.5.

15. The y-axis is a scaled version of the p-value, or the probability that such a result would occur by chance from random sampling variation. The negative log-10 scale is used to make the smaller probabilities appear to be "higher" on the chart and to fit neatly. The dotted line represents the genome-wide p-value threshold of 5×10^{-8}.

16. When the phenotype or other particular requirements for a study are found only in one sample, a less desirable approach that researchers often deploy is to randomly split their sample into two halves and show that the effects manifest in both subgroups.

17. The winner's curse results from the selection of the best SNPs in the discovery sample, which shifts the distribution of effects relative to the replication sample, which may result in regression to the mean. See, for example, L. Xu, R. V. Craiu, and L. Sun, "Bayesian methods to overcome the winner's curse in genetic studies," *Annals of Applied Statistics* 5, no. 1 (2011): 201–231.

18. One important exception to this pattern of disappointment is provided by variation in the gene *FTO*, which has very robustly been shown to predict body mass index and a range of intermediate behaviors that, in turn, affect girth.

19. T. A. Manolio, F. S. Collins, N. J. Cox, D. B. Goldstein, L. A. Hindorff, D. J. Hunter, M. I. McCarthy, et al., "Finding the missing heritability of complex diseases," *Nature* 461, no. 7265 (2009): 747–753.

20. B. Maher, "Personal genomes: The case of the missing heritability," *Nature News* 456, no. 7218 (2008): 18–21.

21. T. A. Manolio, et al., "Finding the missing heritability."

22. For a more complete list see "Dominant and recessive characteristics," at http://www .blinn.edu/socialscience/ldthomas/feldman/handouts/0203hand.htm. If, for example, having one G and one A at a given locus (heterozygosity) has no effect (compared to having G-G) but having A-A has a huge effect, the additive heritability model may be a poor approximation. The same holds true for other nonlinear functions such as both homozygotes GG and AA having the same phenotype and the heterozygote GA being the different one. If most of what we saw in the population were these kinds of nonlinear, or multiplicative, effects, then a simple additive model would miss them. (It should be said that there are both twin and other models for heritability that allow for nonlinearities.)

23. T. D. Howard, G. H. Koppelman, J. Xu, S. L. Zheng, D. S. Postma, D. A. Meyers, and E. R. Bleecker, "Gene-gene interaction in asthma: IL4RA and IL13 in a Dutch population with asthma," *American Journal of Human Genetics* 70, no. 1 (2002): 230–236.

24. Such scenarios, in which the overall additive heritability as estimated by relatedness (such as in the ACE model) were inflated by gene-gene interactions, have been shown in several cases. O. Zuk, E. Hechter, S. R. Sunyaev, and E. S. Lander, "The mystery of missing heritability: Genetic interactions create phantom heritability," *Proceedings of the National Academy of Sciences* 109, no. 4 (2012): 1193–1198.

25. That is, assuming each allele was split 50–50 in the population and parents randomly mated.

26. We can and do accumulate and store neutral (cryptic) mutations across the genome, and then, when they meet in a descendant, they could become "activated" and lead to good (or, more likely, bad) things, depending on the shifting environmental landscape. There are occasions (called decanalization) when previously neutral or insignificant mutations cross a threshold and lead to noticeable phenotypic changes. This is one current theory on the origins of modern chronic diseases, such as schizophrenia, obesity, and diabetes: the environment has shifted just enough to make a bunch of accumulated mutations matter in ways they did not before, so suddenly we see the emergence of a "genetic" disease that did not present in the population previously. Thus, decanalization can explain the emergence of novel phenotypes, but it is unlikely to explain why siblings resemble each other in the cross-section (i.e., heritability). This process of decanalization was first described in a series of famous experiments by Conrad Hal Waddington. Basically, he would expose fruit flies to a shock of some sort during development (one example was raising the temperature at a critical point). Then he would observe the phenotypic change— such as the growth of an additional pair of wings. Then he would select the siblings of those who were exposed and showed the phenotypic response. He would breed those siblings (who were never exposed themselves) and then do it again. After a few generations, the offspring of the selected siblings were sprouting an extra pair of wings without any developmental shock to their environment. And this was in a line of individuals who were never themselves exposed to the novel environment. Rather, it was indirect selection on the siblings at each generation. In other words, the cryptic genetic variation was present all along and was decanalized (or derepressed) by selection on the novel environmental shock. C. H. Waddington, "Canalization of development and the inheritance of acquired characters," *Nature* 150 (1942): 563–565, and "Genetic assimilation of an acquired character," *Evolution* 7 (1953): 118–126.

27. Presumably, the giraffe started out proportioned like a horse (its actual closest living relative is the okapi, which does indeed have the body shape of a horse). To reach the higher leaves, some of the more intrepid of these creatures stretched their necks. Over time, they managed to extend their necks even in a resting state. When they then reproduced, the

necks of their offspring might not be as long as they had stretched their own over a lifetime, but they would be longer than the offspring of those who had been content with browsing the lower-level shoots and leaves. Over the course of generations on this path to higher and higher food sources, you eventually get the giraffe, built on the cumulative efforts of its ancestors.

28. A. Okbay, J. P. Beauchamp, M. A. Fontana, J. J. Lee, T. H. Pers, C. A. Rietveld, P. Turley, et al., "Genome-wide association study identifies 74 loci associated with educational attainment," *Nature* 533, no. 7604 (2016): 539–542.

29. A. R. Wood, T. Esko, J. Yang, S. Vendantam, T. H. Pers, S. Gustafsson, A. Y. Chu, et al., "Defining the role of common variation in the genomic and biological architecture of adult human height," *Nature Genetics* 46 (2014): 1173–1186, http://www.nature.com/ng/journal/v46/n11/full/ng.3097.html.

30. N. Chatterjee, B. Wheeler, J. Sampson, P. Hartge, S. J. Chanock, and J.-H. Park, "Projecting the performance of risk prediction based on polygenic analyses of genome-wide association studies," *Nature Genetics* 45, no. 4 (2013): 400–405; F. Dudbridge, "Power and predictive accuracy of polygenic risk scores," *PLoS Genetics* 9, no. 3 (2013): e1003348.

31. H. D. Daetwyler, B. Villanueva, and J. A. Woolliams, "Accuracy of predicting the genetic risk of disease using a genome-wide approach," *PLoS One* 3, no. 10 (2008): e3395.

32. Imputation can help. Imputation takes the pattern of measured SNPs and matches it with reference genomes that have been fully sequenced (i.e., almost every base pair with a notable amount of variation in the population has been called) by the Human Genome Diversity Project. Based on a probability of those SNPs in between the measured SNPs matching one of the sequenced, reference genomes, it then assigns a probabilistic SNP value for the locus that has not been measured in your data. This is possible due to linkage disequilibrium (LD). LD variants are inherited in small chunks along the chromosome, and the probability of a given SNP being of a certain variety is correlated with those around it, because it is a fellow traveler. The more reference genomes sequenced, the more densely genotyped the SNP chip is and the less variation in the given population (due to population bottlenecks or drift), the greater the accuracy of the imputation. Imputation is directly relevant to the issue of race, as we will discuss in appendix 6. But for present purposes, it is important to remember that as good as it is getting these days, imputation is not the same as measuring each locus for each person.

33. Kaiser Permanente Research Program on Genes, Environment, and Health, Division of Research, Kaiser Permanente, https://www.dor.kaiser.org/external/DORExternal/rpgeh/index.aspx.

34. D. Conley, B. W. Domingue, D. Cesarini, C. Dawes, C. A. Rietveld, and J. D. Boardman, "Is the effect of parental education on offspring biased or moderated by genotype?" *Sociological Science* 2 (2015): 82–105.

CHAPTER 4: GENETIC SORTING AND CAVORTING IN AMERICAN SOCIETY

1. A year later, in 1959, the sociologists Seymour Lipset and Reinhard Bendix noted: "Widespread social mobility has been a concomitant of industrialization and a basic characteristic of modern industrialized society. In every industrialized country, a large proportion of the population have had to find occupations considerably different from those of their parents" (S. M. Lipset and R. Bendix, *Social Mobility in Industrial Societies* [Berkeley: Uni-

versity of California Press, 1959], p. 11). Or in the words of Sibley as paraphrased by Blau and Duncan: "The achieved status of a man, what he has accomplished in terms of some objective criteria, becomes more important than his ascribed status, who he is in the sense of what family he comes from" (P. M. Blau and O. D. Duncan, *The American Occupational Structure* [New York: John Wiley, 1967], p. 430). That is, the notion of a rising meritocracy was not just a joke among social scientists.

2.   They do not cite Young, which is an amazing omission.

3.   Many critics pointed out the flaws of their intriguing argument at the time—particularly sociologists (see, e.g., C. S. Fischer, M. Hout, M. Sanchez Jankowski, S. R. Lucas, A. Swidler, and K. Voss, *Inequality by Design* [Berkeley: University of California Press, 1996]). For example, Herrnstein and Murray (*The Bell Curve: Intelligence and Class Structure in American Life* [New York: Free Press, 1994]) were said to have overestimated the influence of genes on IQ and, in turn, of IQ on socioeconomic attainment. They assumed that an effect of IQ was the effect of genetic endowment, neglecting the vast literature (including that of behavioral genetics) that demonstrated that IQ is subject to large environmental influences and that even its heritability is contingent upon family socioeconomic status. While the criticisms were many, it is worth pointing out that even the second, more controversial part of Herrnstein and Murray's argument was not as far out of the sociological mainstream—namely, the functionalist tradition—as many sociologists would like to believe. Contrast, for instance, the following three quotations:

  1. "For, if there is nothing which either unduly hinders or favors the chances of those competing for occupations, it is inevitable that only those who are most capable at each type of activity will move into it. The only factor which then determines the manner in which work is divided is the diversity of capacities ...."

  2. "Social inequality is thus an unconsciously evolved device by which societies insure that the most important positions are conscientiously filled by the most qualified persons."

  3. "No one decreed that occupations should sort us out by our cognitive abilities, and no one enforces the process. It goes on beneath the surface, guided by its own invisible hand."

  These quotations were penned almost exactly fifty years apart. The first is from Durkheim (1893; in *Emile Durkheim: Selected Writings*, ed. A. Giddens [Cambridge: Cambridge University Press, 1972], p. 181); the second from K. Davis and W. E. More, "Some principles of stratification," *American Sociological Review* 10, no. 2 (1945): 242–249; and the third is from *The Bell Curve* (Herrnstein and Murray, p. 52). With the exception of the fact that the term "cognitive abilities" was not in widespread use in the nineteenth or mid-twentieth centuries, we argue that these quotations could be interchangeable.

4.   Herrnstein and Murray, *The Bell Curve*, 109–110.

5.   Ibid., 91–92.

6.   Ibid., 341.

7.   B. W. Domingue, D. Conley, J. Fletcher, and J. D. Boardman, "Cohort effects in the genetic influence on smoking," *Behavior Genetics* 46, no. 1 (2016): 31–42.

8.   This pattern is also obtained using the polygenic score approach, which we will discuss more below. Our results build upon an important earlier paper that used the twin-based approach to map heritability by birth cohort and found similar results: J. D. Boardman, C. L. Blalock, and F. C. Pampel, "Trends in the genetic influences on smoking," *Journal of Health and Social Behavior* 51, no. 1 (2010): 108–123.

9.   Think about how American-born children of immigrant parents—when the parents come from a poor country—often tower over their elder family members. A dramatic example of this phenomenon can be seen in the case of Ethiopian Jews who were air-lifted out of an extreme low-calorie environment to the land of milk and honey (Israel). Those who were in utero during the migration ended up significantly taller than those who were

already born by the time the Operation Moses evacuation from Addis Ababa occurred in 1984. V. Lavy, A. Schlosser, and A. Shany, "Out of Africa: Human capital consequences of in utero conditions" (working paper w21894, National Bureau of Economic Research, Cambridge, MA, 2016).

10. We see the same pattern if we look at the overall additive heritability of height or body mass index (BMI) and split according to birth cohort. Though heritability increases, we are not adequately powered to detect a statistically significant difference between the two birth cohort groups. D. Conley, T. Laidley, D. W. Belsky, J. M. Fletcher, J. D. Boardman, and B. W. Domingue, "Assortative mating and differential fertility by phenotype and genotype across the 20th century," *Proceedings of the National Academy of Sciences* (2016): 201523592.

11. See: D. Conley, T. M. Laidley, J. D. Boardman, and B. W. Domingue, 2016. Changing Polygenic Penetrance on Phenotypes in the 20th Century Among Adults in the US Population. *Scientific Reports*, 6. One potential concern with this exercise is mortality bias. Since Health and Retirement Study (HRS) respondents had to survive until the 2000s in order to be included in the sample with genotype information, the earlier cohorts in our analysis include individuals who lived longer on average. If individuals with low values of the underlying pro-education genotype (as measured by the polygenic score) and high levels of schooling (or the converse) died earlier, we might observe a stronger correlation between genotype and education in the older cohorts because of differential mortality. This is highly unlikely since a vast research literature suggests that education is positively related to longevity (see, e.g., A. Lleras-Muney, "The relationship between education and adult mortality in the United States," *Review of Economic Studies* 72, no. 1 [2005]: 189–221). If the polygenic score exerts a similar positive effect on longevity (independent of education), then individuals with low educational attainment and low values of the polygenic score would face the highest mortality rates. In that case, selection bias would work to attenuate the relationship between the polygenic score and education. Since this selection bias would be greater for the older cohorts, it would cause us to *underestimate* the decline in the importance of genetic factors. The approach deployed to generate the polygenic score could also bias our results.

On average, the discovery sample (for estimating the weights that are used to make the polygenic score) consists of individuals belonging to younger birth cohorts than those in the Swedish Twin Registry. The mean year of birth for all the cohorts in the discovery sample is 1951.3, whereas the mean year of birth for those included in our HRS analysis is 1941. Even if genetic factors as a whole are similarly important across birth cohorts, if *different* genetic factors matter for different birth cohorts, then a polygenic score constructed with weights from younger birth cohorts may have worse predictive power in older birth cohorts. Since our polygenic score is estimated from younger cohorts, on average, then this source of bias would cause it to predict education better for the younger cohorts in the HRS. Thus, this source of bias is also unlikely to drive the pattern we show above.

Finally, as a check we ran the same exercise with the Framingham Heart Study sample generations 2 and 3, where the median age at the time of genotyping was 39; thus, mortality bias would not likely be an issue. We found no significant change in the interaction between birth cohort and the predictive power of the polygenic score in that sample, either (though with a sample size that is about one-third that of HRS, it was also underpowered).

12. GCTA heritability analysis shows no change between younger and older birth cohort groupings but suffers from the same problem of inadequate statistical power as did the analyses for height and BMI.

13. See, for example, A. R. Branigan, K. J. McCallum, and J. Freese, "Variation in the heritability of educational attainment: An international meta-analysis," *Social Forces* 92 (2013):109–140. Their meta-analysis of twin studies across the globe evinced a pattern whereby "For men and individuals born in the latter half of the twentieth century, more of the variance in attainment can be explained by genetic variation" (p. 132). However, they base this claim on differences across samples (and, for that matter, nations) rather than on differences within a given population and sample. Furthermore, their observed differences in the partitioning of variance could be due to the changing nature of twinning (more twins born to older couples due to rising age at first birth) and/or differences in prenatal conditions, which have been shown to affect twin-based heritability estimates (see, e.g., B. Devlin, M. Daniels, and K. Roeder., "The heritability of IQ," *Nature* 388, no. 6641 (1997): 468–471).

14. In order to pinpoint where the decline in the genetic effect of education was occurring in the distribution, we conducted analyses where we estimated linear probability models for educational transitions (see R. D. Mare, "Social background and school continuation decisions," *Journal of the American Statistical Association* 75, no. 370 [1980]: 295–305.). We focused on at least finishing high school (≥12 years of attainment), at least some college (>12 years of attainment), finishing college or more (≥16 years of attainment), or more than college (>16 years of attainment). For each educational stage our analyses focused on those who had completed the step below (i.e., we looked at those who had attended at least some college among those who had only graduated from high school). When we examined the strength of the educational polygenic score as a predictor across educational transitions (i.e., a main effect constrained to be constant across birth cohorts), we found that it increased as we moved from high school completion to the transition to and completion of college. However, when we modeled the transition from college to graduate school, it once again declined. We then examined the interaction between the polygenic score and birth year and found that it was at the lower end of the distribution (high school graduation), where the effect of the polygenic score was declining. Indeed, at the highest educational transition—from college completion to graduate school—we found that the effect of genetics is increasing in younger birth cohorts. We also tested these models using a polygenic score from the same consortium (C. A. Rietveld, S. E. Medland, J. Derringer, J. Yang, T. Esko, N. W. Martin, H.-J. Westra, et al., "GWAS of 126,559 individuals identifies genetic variants associated with educational attainment," *Science* 340, no. 6139 [2013]: 1467–1471) specifically meant to predict college completion and got the same result (D. Conley and B. Domingue, "The Bell Curve revisited: Testing controversial hypotheses with molecular genetic data," *Sociological Science* 3 [2016], doi:10.15195/v3.a23).

The educational stage analysis speaks to debates over the impact of family background within the education system. Mare ("Social background") suggests that the further up the educational "ladder" a given transition lies, the less family background should matter due to selection gradients at each stage. While educational polygenic score is part of one's patrimony, it demonstrates significant variation within families and, indeed, is arguably a measure of the skill-based endowment on which the selection gradient of universalism should operate. We found that across all birth cohorts, the effect of the educational polygenic score is largest for college attainment and weaker for high school completion and graduate transitions—thus demonstrating an inverse U-shaped pattern that does not conform to the selection gradient theory. Deploying this educational stage analysis, we also tested a genetic version of the maximally maintained inequality theory (MMI; see A. E. Raftery and M. Hout, "Maximally maintained inequality: Expansion, reform, and opportunity in Irish education, 1921–75," *Sociology of Education* [1993]: 41–62). A prediction of the MMI theory is that as a given educational stage (such as secondary school) approaches saturation (i.e., is universally accessible), class background should matter less for that stage. As secondary schooling became universally accessible, we should have seen the decline of the polygenic score effect for high school graduation over time, but since post-secondary schooling did not expand to the same inequality-dampening degree, for transitions within the post-secondary system we expected no decline in the impact of the polygenic score in younger birth cohorts. Here, in line with theories of maximally maintained inequality, we found that the decline in the importance of the genetic profile is seen in the lower half of the educational distribution (primarily high school completion and transition to post-secondary schooling). Indeed, we find that for transition from a completed college degree to graduate education, the trend of the effect of the polygenic score is positive over time (Conley and Domingue, "Bell Curve revisited").

15. Of course, much of the change in education levels over this time period in the United States is happening beyond the "mandatory" requirements (typically 10 years of schooling), with the expansion of higher education in the second half of the century, which reasserts the possibility that genotype should matter.

16. A. Okbay, J. P. Beauchamp, M. A. Fontana, J. J. Lee, T. H. Pers, C. A. Rietveld, P. Turley, et al., "Genome-wide association study identifies 74 loci associated with educational attainment," *Nature* 533, no. 7604 (2016): 539–542.

17. D. Conley and B. Domingue, "Bell Curve revisited."

18. Of course, while education and, to a lesser extent, occupational choice, are often (but not always) determined prior to the forming of relationships, earnings are highly endogenous to decisions after relationships are formed. For a review of the assortative mating literature, see C. Schwartz, "Trends and variation in assortative mating: Causes and consequences," *Annual Review of Sociology* 39 (2013): 451–470.

19. C. R. Schwartz and R. D. Mare, "Trends in educational assortative mating from 1940 to 2003," *Demography* 42, no. 4 (2005): 621–646.

20. Perhaps we could even witness a greater rise in genotypic sorting than in phenotypic sorting, since "college graduate" is a pretty big basket that obscures a lot of subtle differences that may be reflected in genotypic similarity. In other words, genotype may be a better measure of what spouses are looking for (cognitive ability, ability to delay gratification, and other traits that lead to academic success) than simply the presence or absence of a B.A. or a B.S. degree.

21. J. R. Alford, P. K. Hatemi, J. R. Hibbing, N. G. Martin, and L. J. Eaves, "The politics of mate choice," *Journal of Politics* 73, no. 2 (2011): 362–379.

22. See, for example, C. T. Gualtieri, "Husband-wife correlations in neurocognitive test performance," *Psychology* 4, no. 10 (2013): 771–775.

23. Ibid.

24. Scholars checking couples at different points during their marriage have not found increased similarity, thus potentially ruling out convergence. However, studying individuals who are already married makes it difficult to distinguish between possible explanations. Ideally, we would also collect information about couples before they were married. See, for example, M. N. Humbad, M. B. Donnellan, W. G. Iacono, M. McGue, and S. A. Burt, "Is spousal similarity for personality a matter of convergence or selection?" *Personality and Individual Differences* 49, no. 7 (2010): 827–830; R. R. McCrae, T. A. Martin, M. Hrebícková, T. Urbánek, D. I. Boomsma, G. Willemsen, and P. T. Costa Jr., "Personality trait similarity between spouses in four cultures," *Journal of Personality* 76, no. 5 (2008): 1137–1164; D. Watson, E. C. Klohnen, A. Casillas, E. N. Simms, and J. Haig, "Match makers and deal breakers: Analyses of assortative mating in newlywed couples," *Journal of Personality* 72,

no. 5 (2004): 1029–1068; G. C. Gonzaga, S. Carter, and J. G. Buckwalter, "Assortative mating, convergence, and satisfaction in married couples," *Personal Relationships* 17, no. 4 (2010): 634–644.

25. In two different studies, political scientists analyzed the content of *actual* dating profiles on a popular online dating site. They found that individuals were reluctant to share political information about themselves in their profiles. However, individuals may make dating choices based on nonpolitical characteristics that tend to be shared among liberals and conservatives. Thus, while the choice of a mate is not explicitly based on political ideology, this dynamic can still lead to assortative mating. See C. A. Klofstad, R. McDermott, and P. K. Hatemi, "Do bedroom eyes wear political glasses? The role of politics in human mate attraction," *Evolution and Human Behavior* 33, no. 2 (2012): 100–108 and "The dating preferences of liberals and conservatives," *Political Behavior* 35, no. 3 (2013): 519–538.

26. A recent study by a team of political scientists suggests an intriguing mechanism that allows individuals to identify potential mates with similar political attitudes—by smell. More specifically, individuals may detect olfactory cues from like-minded individuals. The logic behind the study is that for a number of evolutionary reasons, such as disease avoidance and cheater detection, it is useful to identify those belonging to in-groups and out-groups. Olfactory information tends to aid these efforts. The researchers conducted an experiment that asked participants to rate the attractiveness of the body odor of anonymous individuals (targets) on the far left (liberal) or far right (conservative) on the political spectrum. The experimental subjects tended to rate the odor of targets with similar political leanings more favorably than those with opposing views. While provocative, more research must be conducted before the link between assortative mating on political attitudes and olfactory clues is firmly established. An excellent example of this is the disgust response of humans when smelling rotten food. This physical response serves to protect us from eating something that will make us ill or, worse, kill us. Researchers have found that disgust sensitivity is highly correlated with conservative attitudes, especially in the areas of morality and sexuality. R. McDermott, D. Tingley, and P. K. Hatemi, "Assortative mating on ideology could operate through olfactory cues," *American Journal of Political Science* 58, no. 4 (2014): 997–1005.

27. L. Eika, M. Mogstad, and B. Zafar, "Educational assortative mating and household income inequality" (working paper w20271, National Bureau of Economic Research, Cambridge, MA, 2014), http://www.newyorkfed.org/research/staff_reports/sr682.pdf.

28. R. Breen and L. Salazar, "Educational assortative mating and earnings inequality in the United States," *American Journal of Sociology* 117 (2011): 808–843.

29. Another theory posits that when income inequality is high, there is more social distance between groups and thus individuals are more likely to encounter dating partners within their own class, and further, that when inequality is higher there is more incentive not to marry down, reducing heterogeneity within marriages. R. Fernandez, N. Guner, and J. Knowles, "Love and money: A theoretical and empirical analysis of household sorting and inequality," *Quarterly Journal of Economics* 120 (2005): 273–344.

30. http://discovermagazine.com/2003/aug/featkiss#.UuiJgWQo46g.

31. J. H. Fowler, J. E. Settle, and N. A. Christakis, "Correlated genotypes in friendship networks," *Proceedings of the National Academy of Sciences* 108, no. 5 (2011): 1993–1997. We have also done work in this area by extending the analysis of correlations between the genotypes of friends to show that schools (and, more generally, environments) shape the way these correlations are created. There is a role for social structure to produce correlated genotypes among friends, even if people do not actively seek out similar friends. J. D. Boardman, B. W. Domingue, and J. M. Fletcher, "How social and genetic factors

predict friendship networks," *Proceedings of the National Academy of Sciences* 109, no. 43 (2012): 17377–17381.

32. Perhaps it is because in moths linoleic acid is metabolized into a pheromone. But it is also related to a whole host of bodily functions, including fat-cell function and bone formation. So it could be that friends sort along physical dimensions (a well-known phenomenon).

33. Other cases in which the evolutionary fitness of a given genotype is contingent on the overall distribution of genotypes include negative frequency dependent selection (balancing selection). It is easy to understand this concept if we think of a field of wildflowers. If all the flowers are purple and I experience a mutation that makes me green, I may stand out to and attract the species' pollinators (assuming they are not color-blind). I will thus propagate at a higher rate than those boring purple relatives around me, so the proportion of green will rise in the field over the course of generations. But then when green is the dominant color, purple may be the one that stands out, and thus the advantage shifts, so that the actual distribution of petal hue balances out. This is an example of what geneticists call negative frequency dependent selection. Of course, if I seek to camouflage myself to prevent predation, then the fitness of my genotype depends on the genetic profile of the organisms around me—for example, the hue of a leaf on which I spend my life. This yields a positive synergistic evolutionary path—assuming my color helps the leaf in some way, too.

    We looked at this possibility in humans with respect to behavior, focusing on two well-studied genetic variants in the serotonin neural system and the dopamine reward system. We asked whether the effect of our own genotype was dependent on the genotypes of our siblings. In the competition within the nest for parental attention and investment, you might think having a sensitive, demanding genotype is an advantage if your siblings have the more "chill" genotype. Or it could go the other way: if you are the only squeaky wheel in the brood, you get stigmatized and shunned, but if you all have the demanding alleles, you form a union of sorts and extract extra resources from your parents.

    To our surprise, we found that for both the dopamine and the serotonin systems, having the genetic variants that were generally characterized as the more sensitive versions (see chapter 7 for a thorough discussion of this theory) had no effect on your likelihood of depression or how well you did in school (outcomes that may reflect family dynamics) if you were the only one who had that genotype. But when you and your sibling (in this case your fraternal twin) had the same sensitivity, everyone was worse off in terms of risk for suboptimal outcomes. Of course, as was discussed in earlier chapters, the fact that this was a candidate gene study based on a relatively small sample means that the results should be considered suggestive only if they are replicated.

34. It could be that the height polygenic score has other outward signs of pleiotropic effects—such as general haleness—upon which potential mates sort. This is an intriguing mystery.

35. We should also note that the positive genetic correlation between parents on most of the traits suggests that classical twin-based approaches to estimating heritability are biased downward. Recall from chapter 2 that the ACE model assumes that identical twins are twice as genetically related as fraternal twins; hence the doubling of the difference in sibling correlations between identicals and same-sex fraternals to obtain A (the additive genetic component). To the extent that parents are more similar genetically (on the causal alleles), fraternal twins are also more than 50 percent genetically similar on average (for the alleles that matter), so the difference in correlations should be multiplied by a factor greater than 2. Of course, this source of bias may be offset by others.

36. Indeed, when we remove the component of education that is predicted by the education polygenic score and look at the spousal correlation in the residual—the part not predicted by the polygenic score—it hardly budges. Thus, from the height of the bars in figure 4.5,

it looks like the genotypic sorting is about one-fourth as strong as the phenotypic sorting, but they are really two different processes.

37. It is possible that there is an increase in genetic sorting on education, but this particular measure does not indicate that to be the case. For that to be true, it would require heroic assumptions about how unrepresentative the polygenic score is of the overall genetic contribution to education or IQ. A second problem could be related to survival bias. We limited our analysis to first marriages, so it could be the case that those with more similar genetic scores are more likely to end up staying married longer (or, for that matter, living longer) than those couples who have more divergent genotypes. If that were the case, we could observe this cohort difference due to differential sample selection. Namely, the older group would appear to have more similar genotypes because they are the ones who stayed alive and married long enough to be in the sample, and their compatriots with more disparate scores who would have been in the sample are either dead or divorced. While we cannot assess the death hypothesis directly (we do discuss and dismiss its relevance to the overall genotype-phenotype trend above), we can see if our results change if we allow second- and higher-order marriages to be included in the group. Adding these couples to the sample does not change the overall trends.

38. Now that we have dispensed with this marriage-related *Bell Curve* proposition, we would really like to know the answer to a related question, namely, whether the combined genetics of you and your spouse—such as discordances on particular polygenic scores—predict the quality of a marriage and thus the likelihood that you will break up. The available data, however, are not yet up to this task. In thinking about this question within the American context, there are two major datasets in the United States that have genotypic information on both spouses. As mentioned above, spouses in the Health and Retirement Study are too old by the time we observe them (the minimum age is 50), so the ones who die or divorce beforehand are not observed (though we may see their second or third marriages). And the Framingham Heart Study second-generation sample, which includes spouses, simply does not have enough cases to track their marriages prospectively by genotype matches. Better still would be to have a registry of all single individuals from the time they enter the dating market to observe how their genotypes influence relationship formation. Unfortunately, the data currently available does not allow us to address such questions. To tackle the relationship formation question, we would need to genotype a population, which the Scandinavian countries—with their extensive national registries—could feasibly do (but have not yet). The United States has strong norms against such comprehensive data collection.

39. See, for example, R. Lynn, *Dysgenics: Genetic Deterioration in Modern Populations* (Santa Barbara, CA: Praeger Publishers, 2011).

40. T. J. Mathews and S. J. Ventura, "Birth and fertility rates by educational attainment: United States, *Monthly Vital Statistics Report* (National Center for Health Statistics) 45, no. 10 suppl. (1994): 97–1120.

41. R. D. Mare and V. Maralani, "The intergenerational effects of changes in women's educational attainments," *American Sociological Review* 71 (2006): 542–564.

42. See, for example, F. C. Tropf, R. M. Verweij, P. J. van der Most, G. Stulp, A. Bakshi, D. A. Briley, M. Robinson, et al., "Mega-analysis of 31,396 individuals from 6 countries uncovers strong gene-environment interaction for human fertility," *bioRxiv* (2016): 049163.

43. The caveats regarding potential mortality bias apply here as well.

44. This figure shows the Manhattan plot from the *Science* paper that created the first polygenic score for educational attainment. If the genetic forces that predicted education were the same for women and men, the buildings in the Manhattan plot should align closely,

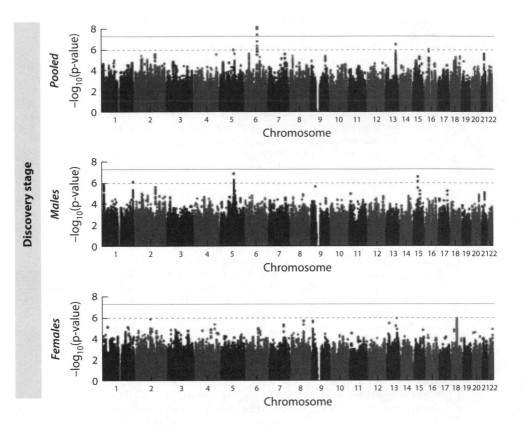

Figure 4.7. Manhattan plots of SNPs for EduYears in single genomic control meta-analysis. SNPs are plotted on the *x*-axis according to their position on each chromosome against association with EduYears on the *y*-axis (shown as $-\log^{10}$ *p*-value). The solid line indicates the threshold for genome-wide significance ($p<5\text{x}10^{-8}$) and the dashed line the threshold for suggestive hits ($p<1\text{x}10^{-6}$).

which is not the case. Although there are no statistically detectable differences in many of the genetic effects, these results could, at the least, suggest that further investigation of gender differences is necessary.

Of course, in order to conduct these analyses of gender differences, we need ever-larger sample sizes. This explanation is of interest because it suggests environmental processes that might filter women and men into college in different ways.

CHAPTER 5: IS RACE GENETIC?

1. F. González-Andrade, D. Sánchez, J. González-Solórzano, S. Gascón, and B. Martínez-Jarreta, "Sex-specific genetic admixture of Mestizos, Amerindian Kichwas, and Afro-Ecuadorans from Ecuador," *Human Biology* 79, no. 1 (2007): 51–77.

2. Genetic drift denotes the change in the frequency of a gene variant in a population due to a random sampling of organisms. The genetic variants in the offspring are a sample of those in the parents, so that (random) chance has a role in determining whether a given genotype survives in a population.

3. R. Bhopal, "The beautiful skull and Blumenbach's errors: The birth of the scientific concept of race," *BMJ* 335 (2007): 1308.

4. He also dismissed skin color and pigmentation as a means of categorization and suggested that the differences across races were mainly driven by climate, which influenced skin color (as well as the hues of plants), skull structure, and stature. Other claims of note made by Blumenbach include that the most beautiful people are from present-day Georgia, and white skin color was the first color in humans (his reasoning being that is easier for skin to be tanned from white to brown than the reverse).

    This classification system has shrunk and also expanded over time. Georges Cuvier (1769–1832) claimed that there were three races—Caucausian, Ethiopian, and Mongolian (J. P. Jackson and N. M. Weidman, *Race, Racism, and Science: Social Impact and Interaction* [New Brunswick, NJ: Rutgers University Press, 2005], 41–42). Carl Linnaeus subdivided the human species into white European, red American, brown Asian, and black African in his *Systema Naturae*, last published in 1793. Rather than holding to a specific number of races, the common denominator among most classifications was the claim of white superiority and black inferiority.

5. We are oversimplifying through omission here. We still have to demarcate the geography (usually by modern country configurations) in a time period. Ancestry is meant to conform to a "native people" idea, but being native is anchored in some time period. English ancestry refers to the population post–Norman Conquest but not to children born in England of "African" parents. And there is no such measure as "USA" ancestry, even though a lineage could now have inhabited a specific area for hundreds of years.

6. One could argue that Census racial categories typically used by social scientists make more genetic sense because we are separating ourselves into three broad groups that seem to diverge from each other on the tree: those (primarily) of African descent, those of Eurasian descent, and Native Americans—reverting from Blumenbach's four or five categories to three. But doing so disregards the high degree of intermarriage between groups (what geneticists call "admixture"). And it totally ignores Latinos, who are the most ad-mixed population, showing nontrivial proportions of Native American, African, and European ancestry, which vary by ethnic group. For example, Mexican-Americans have a high degree of Asian ancestry while Puerto Ricans display a significant level of African descent. (We discuss admixture in more depth later in this chapter.)

7. L. Wacquant, "Deadly symbiosis when ghetto and prison meet and mesh," *Punishment & Society* 3, no. 1 (2001): 95–133.

8. In recent years though, DNA analysis has facilitated many African Americans' ability to reinstantiate ethnic identity by tracing specific lineages via ancestral analysis. See, for example, A. Nelson, *The Social Life of DNA: Race, Reparations, and Reconciliation after the Genome*, (Boston, MA: Beacon, 2016).

9. Kwanzaa being the exception that proves the rule, since it was created only fifty years ago in the United States. Between 1 and 5 percent of African Americans in the United States

report celebrating it. M. Scott, "Kwanzaa celebrations continue, but boom is over," *Buffalo News*, December 20, 2009, http://web.archive.org/web/20091220052310/http://www.buffalonews.com/260/story/897568.html.

10. Among Europeans or those of European descent, for example, 1–3 percent of the genome is of Neanderthal ancestry as a result of interbreeding between the species. See, for example, S. Sankararaman, S. Mallick, M. Dannemann, K. Prüfer, J. Kelso, S. Pääbo, N. Patterson, and D. Reich, "The genomic landscape of Neanderthal ancestry in present-day humans, *Nature* 507 no. 7492 (2014): 354–357.

11. Of course, if we go back further, all humans descend from the primordial "Adam" and "Eve," the putatively first modern humans (though whether there is actually one set of these universal grandparents is a highly debatable proposition). However, within the cradle of Africa, evolution worked for many millennia to allow mutations to accumulate, causing a great degree of genetic variation in the populations that lived—then and now—in the cradle of human origins. Those who left northeastern Africa, however, took with them only the genetic polymorphisms that they happen to have had at the time, a small subset of all the contemporaneous variation in the human species. Of course, new mutations have arisen in the 60,000 years since some humans left the Rift Valley, but these have occurred at the same rate inside and outside of Africa.

12. The neutral theory posits how genetic variation occurs and is maintained through processes of mutation and drift without selective pressures. M. Kimura, *The Neutral Theory of Molecular Evolution* (Cambridge: Cambridge University Press, 1984).

13. More precisely, assuming the alleles are in Hardy-Weinberg equilibrium. Hardy-Weinberg equilibrium refers to the idealized situation in which allele frequencies are stable in a population thanks to a lack of migration, selection, and mutation and the presence of a large effective mating population size in which random mating occurs.

14. If we formally plot another measure of diversity—called the F-statistic—against migratory distance from East Africa, we see this robust pattern (see, e.g., Q. Ashraf and O. Galor, "The 'Out of Africa' hypothesis, human genetic diversity, and comparative economic development," *American Economic Review* 103, no. 1 (2013): 1–46). The F-statistic is a measure of inbreeding that reflects population structure; it is essentially the difference between heterozygosity from what would be expected from allele frequencies in Hardy-Weinberg equilibrium and the observed rate of heterozygosity in a population or across populations.

15. This distinction remains true even after we take into consideration the fact that—as mentioned earlier—those who left Africa actually mated with nonhumans: Neanderthals and Denovians, who made the migration out of Africa before modern humans evolved.

16. While these figures clearly illustrate that African origin versus non-African origin is a key part of the story, you might be wondering how we have data to create these figures. There are indeed criticisms of how the genetic clustering of populations has been conducted. Perhaps the most relevant one is that there is ascertainment bias based on the samples that researchers have obtained for the Human Genome Diversity Project (HGDP) and the HapMap Project. We are attempting to describe processes of migration and genetic drift that took place across tens of thousands of years, and we do not have genetic data for that time period. If we want an accurate representation of the current global genetic landscape, how should we go about it?

The typical approach by researchers is to track down present-day members of self-identified ethnic groups and collect their DNA. This approach rests on a number of assumptions. First, it assumes that within a given geographic range, the people who live there today are the descendants of those in that same group living in that same region from thousands of years ago. (Obviously, when ethnic groups are known to have relocated in recent history that is taken into account.) It also assumes that these self-identified groups form relatively independent mating populations. That is, while the approach does not discount that there is significant genetic variation within a country as large as China or as small as the Netherlands, if we were to take lots of subsamples from different areas of Holland or the People's Republic, it would not change the overall landscape at the global level—namely, figure 5.4 would look the same. So, while an approach that randomly samples DNA across the entire distribution of human populations as they now appear geographically might be ideal, it is highly unlikely that it would change the basic story told above.

But, with some notable exceptions, the hypothesis-free analysis of genetic data largely confirms the earlier work classifying ancestry and migration patterns based on skull morphology and other skeletal characteristics. This approach is further validated by the correspondence between cladograms (classification trees) of ethnolinguistic variation with those of genetic variation. See N. Creanza, M. Ruhlen, T. J. Pemberton, N. A. Rosenberg, M. W. Feldman, and S. Ramachandran, "A comparison of worldwide phonemic and genetic variation in human populations," *Proceedings of the National Academy of Sciences* 112, no. 5 (2015): 1265–1272.

17. "*Zea mays*," Gramene, http://ensembl.gramene.org/Zea_mays/Info/Index.
18. K. McAuliffe, "They don't make Homo sapiens like they used to," *Discover*, March 2009, http://discovermagazine.com/2009/mar/09-they-dont-make-homo-sapiens-like-they -used-to.
19. M. D. Weight and H. Harpending, "Some uses of models of quantitative genetic selection in social science," *Journal of Biosocial Science* (2016): 1–16, doi: 10.1017/S002193201600002X.
20. J. Hawks, E. T. Wang, G. M. Cochran, H. C. Harpending, and R. K. Moyzis, "Recent acceleration of human adaptive evolution," *Proceedings of the National Academy of Sciences* 104, no. 52 (2007): 20753–20758.
21. Even genetic assortative mating could cause genetic differences of significant magnitudes if it were prevalent enough; however, these would likely result in within-society cleavages. See, for example, H. Harpending and G. Cochran, "Assortative mating, class, and caste," in *The Evolution of Sexuality*, ed. T. K. Shackelford and R. D. Hansen (New York: Springer, 2015), 57–67.
22. E. Milot, F. M. Mayer, D. H. Nussey, M. Boisvert, F. Pelletier, and D. Réale, "Evidence for evolution in response to natural selection in a contemporary human population," *Proceedings of the National Academy of Sciences* 108, no. 41 (2011): 17040–17045, http://www.pnas .org/content/108/41/17040.
23. This mistake is similar to partisan differences in attributing success in life to luck (genetic drift) versus effort and ability (natural selection). See R. H. Frank, *Success and Luck: Good Fortune and the Myth of Meritocracy* (Princeton, NJ: Princeton University Press, 2016).
24. B. Bogin and M. I. Varela-Silva, "Leg length, body proportion, and health: A review with a note on beauty," *International Journal of Environmental Research and Public Health* 7, no. 3 (2010): 1047–1075.
25. M. J. Zuidhof, B. L. Schneider, V. L. Carney, D. R. Korver, and F. E. Robinson. "Growth, efficiency, and yield of commercial broilers from 1957, 1978, and 2005," *Poultry Science* 93, no. 12 (2014): 2970–2982, http://ps.oxfordjournals.org/content/early/2014/09/26/ps.2014 -04291.abstract.
26. These changes cannot even be attributed to selective migration—in which the genetically advantaged flock to areas that are flourishing—like the story of China's rise perhaps can; Shanghai now attains a level of income equal to Italy whereas rural western areas are more like some African countries.
27. J. Kourany, "Should some knowledge be forbidden? The case of cognitive differences research," *Philosophy of Science*, 2016 (in press).
28. J. Novembre, T. Johnson, K. Bryc, Z. Kutalik, A. R. Boyko, A. Auton, A. Indap, et al., "Genes mirror geography within Europe," *Nature* 456, no. 7218 (2008): 98–101.
29. The fact that PCA is an atheoretical statistical exercise rather than a hypothesis-based analysis has both benefits and consequences. One benefit is that the procedure merely

combines and summarizes the available data without the researcher forcing a set of pre-ordained race and ethnic labels onto the groupings. One consequence is that the PCs have no clear and specific interpretation. When the data are measuring genetic variables, the PCs capture regularities in DNA. As we discuss above, key regularities in DNA data reflect histories of migration, genetic drift, and admixture, suggesting that the labeling of PCs as "ancestry" is reasonable. But such labeling is one of many assumptions that supply the foundation to analyses that examine potential links between genotypic measures of ancestry and socioeconomic outcomes.

30. This is a pattern not just in our data. See, for example, F. Zakharia, A. Basu, D. Absher, T. L. Assimes, A. S. Go, M. A. Hlatky, C. Iribarren, et al., "Characterizing the admixed African ancestry of African Americans," *Genome Biology* 10, no. 12 (2009): R141; K. Bryc, A. Auton, M. R. Nelson, J. R. Oksenberg, S. L. Hauser, S. Williams, A. Froment, et al., "Genome-wide patterns of population structure and admixture in West Africans and African Americans," *Proceedings of the National Academy of Sciences* 107, no. 2 (2010): 786–791; R. Yaeger, A. Avila-Bront, K. Abdul, P. C. Nolan, V. R. Grann, M. C. Birchette, S. Choudhry, et al., "Comparing genetic ancestry and self-described race in African Americans born in the United States and in Africa," *Cancer Epidemiology Biomarkers & Prevention* 17, no. 6 (2008): 1329–1338.

31. Novembre et al., "Genes mirror geography."

32. K. Bryc, E. Y. Durand, J. M. Macpherson, D. Reich, and J. L. Mountain, "The genetic ancestry of African Americans, Latinos, and European Americans across the United States," *American Journal of Human Genetics* 96, no. 1 (2015): 37–53.

33. For examples of how to use PCs to map ancestry, see, for example, A. L. Price, N. A. Zaitlen, D. Reich, and N. Patterson, "New approaches to population stratification in genome-wide association studies," *Nature Reviews Genetics* 11, no. 7 (2010): 459–463; G. McVean, "A genealogical interpretation of principal components analysis," *PLoS Genetics* 5, no. 10 (2009): e1000686.

34. S. Baharian, M. Barakatt, C. R. Gignoux, S. Shringarpure, J. Errington, W. J. Blot, C. D. Bustamante, et al., "The Great Migration and African-American genomic diversity," *PLoS Genetics* 12, no. 5 (2016): e1006059.

35. After all, that is their whole point: to cluster by "tribe," and families are the most basic tribe.

36. Due to the randomness of recombination and segregation.

37. In the Framingham Heart Study and the Minnesota Twin Family Study, which are both highly homogeneous samples of almost exclusively white siblings. So these sibling differences in PCs are not detecting what we would call race differences but, instead, ethnic ones. Or rather, something akin to differences in ethnic-proportions—such as how much of each of the sibling's DNA melting pot contains German ingredients compared with Italian.

38. Intuitively, this analysis seems straightforward, but many of the details have not been worked out. The approach would use the idea that each sibling has randomly received a genetic letter at each location, based on the possible letters that his parents provided. The extent that siblings' letters differ at each location is random, so there are millions of little genetic lotteries taking place that we might be able to use to estimate causal effects on outcomes. One detail is that siblings can receive only letters that their parents have—they can be different only in places their parents are heterozygous (have more than one letter), so not all pairs of siblings are "eligible" for every little genetic lottery, and we do not yet know if siblings are, generally, eligible for the lotteries that make up the important parts of a given polygenic score.

A second detail is that the polygenic score adds together SNP information in a very simple way, based on the number or importance of "risky" alleles across the genome. A sibling model would just take the difference in the final polygenic score number between siblings and compare this difference to variation in siblings' educational attainments. But what if this is too simple? What if all the "action" of the education polygenic score is in the first chromosome, and some siblings have almost no differences in that chromosome and others have many? Or what if the source of discordancy in the score for one set of siblings is mostly in chromosome 8 but in others it is in chromosome 6 due to differences in assortative mating patterns? Then on the surface the sibling differences may appear to be random within each family, but population structure may sneak back in, depending on where in the genome those differences in scores arise across sibling pairs.

Third, as mentioned earlier, the sibling difference model—even for candidate gene studies—breaks the linkage between population structure (and thus any cross-family environmental differences) and allele effects; however, while internally consistent for within-family comparisons, it remains an open question as to whether estimates generated within families are externally valid for cross-family differences in the same genotype-phenotype relationship. That is, due to niche formation, or alternatively, compensation within the family, it may be that such models over- or underestimate the effect we want to generalize to the entire population. Indeed, as discussed in chapter 3, the educational polygenic score seems to have bigger effects within families than between them and acts as an engine of mobility in this fashion.

A final detail is one that we have already discussed: PGS and PC analysis hinders us from understanding genetic mechanisms. A score of 11 versus 10 on a polygenic score for education increasing schooling by one month reveals less about specific genetic mechanisms than a candidate gene approach in which we can show that an A rather than a T at a position in a gene increases schooling by one day. Though a fair counterpoint is that we may never have sibling datasets large enough to estimate the effects of genotype on schooling if we are looking for effect sizes of one day.

39. But sibling analysis raises its own issues, such as spillover effects from one to the other, which would mute effects within families, or niche formation, which would overstate them. Plus, identifying effects from discordancy within families raises its own issues when there is sizable measurement error. See, for example, T. Frisell, S. Öberg, R. Kuja-Halkola, and A. Sjölander, "Sibling comparison designs: Bias from non-shared confounders and measurement error," *Epidemiology* 23, no. 5 (2012): 713–720.

40. Though other less plausible explanations still remain possible—such as a gene-environment interaction that leads to different etiologies of birth weight by race—such alternatives are much fewer than in a "naïve" analysis of race differences. See: Dalton Conley and Kate W. Strully. "Birth weight, infant mortality, and race: Twin comparisons and genetic/environmental inputs." *Social Science & Medicine* 75, no. 12 (2012): 2446–2454.

41. Of course, it could still be that healthier individuals have jobs that keep them away from the sun (and thus lighter), but again, the range and plausibility of alternative explanations is greatly attenuated with genetic controls. Under other assumptions, we could use the genes for skin tone to predict (instrument for) for sibling differences in color in order to assess the purely environmental effect of that genetically- (not sun-) induced skin tone on health outcomes. (Here the assumption is that the skin-tone genes do not directly affect blood pressure.)

42. Ann J. Morning, Department of Sociology, New York University, email.

CHAPTER 6: THE WEALTH OF NATIONS: SOMETHING IN OUR GENES?

1. J. d'A. Guedes, T. C. Bestor, D. Carrasco, R. Flad, E. Fosse, M. Herzfeld, C. C. Lamberg-Karlovsky, et al., "Is poverty in our genes?" *Current Anthropology* 54, no. 1 (2013): 71–79.
2. http://data.worldbank.org/topic/poverty.

3. "GDP per capita," Data, World Bank, http://data.worldbank.org/indicator/NY.GDP.PCAP
    .CD?order=wbapi_data_value_2013+wbapi_data_value+wbapi_data_value-last&sort=asc.
4. "Life expectancy at birth, female (years), Data, World Bank, http://data.worldbank.org
    /indicator/SP.DYN.LE00.FE.IN?order=wbapi_data_value_2012+wbapi_data_value+wbapi
    _data_value-last&sort=asc.
5. "Korea, Rep.," Data, World Bank, http://data.worldbank.org/country/korea-republic.
6. "North Korea," The World Factbook, Central Intelligence Agency, https://www.cia.gov
    /library/publications/the-world-factbook/fields/2004.html#kn.
7. Recall, also, that North Korea experienced overwhelming destruction during the Korean
    War in the 1950s. One might wonder whether the devastation, and not necessarily the
    institutions set up under Communist control, is more to blame for the lack of develop-
    mental success in the country. Alternatively, some reports suggest that industrial produc-
    tion in North Korea rebounded completely by 1957. North Korea also had some initial
    advantage because of its faster industrialization under Japanese rule between 1910 and
    1945. Comparisons of differences in outcomes "close" to the thirty-eighth parallel are
    particularly compelling because of the arbitrariness of the dividing line. R. L. Worden,
    ed., *North Korea: A Country Study*, 5th ed. (Washington, DC: Federal Research Division,
    Library of Congress, 2008.
8. African river blindness (onchocerciasis), the second most common cause of blindness
    due to infection (it affects 17–25 million people), is a disease caused by infection from a
    parasitic worm that is spread by a type of black fly that lives near rivers, especially in sub-
    Saharan Africa.
9. S. Enrico and R. Wacziarg, "How deep are the roots of economic development?" *Journal of
    Economic Literature* 51, no. 2 (2013): 325–369.
10. Examples suggested by Jared Diamond are that temperate plants store more energy in
    parts edible to humans than tropical plants and that more recent glacier advance and re-
    treat have created nutrient-rich soils outside the tropics. J. Diamond, "What makes coun-
    tries rich or poor?" review of *Why Nations Fail: The Origins of Power, Prosperity, and Poverty*,
    by Daron Acemoğlu and James Robinson, *New York Review of Books*, June 7, 2012, http://
    www.nybooks.com/articles/2012/06/07/what-makes-countries-rich-or-poor/.
11. Of course, east-west exchange was not always beneficial, as evidenced by the bubonic
    plague killing roughly one-third of the European population during the fourteenth
    century.
12. O. Galor and Ö. Özak, "The agricultural origins of time preference" (working paper
    w20438, National Bureau of Economic Research, Cambridge, MA, 2014).
13. T. Talhelm, A. Zhang, S. Oishi, C. Shimin, D. Duan, X. Lan, and S. Kitayama, "Large-scale
    psychological differences within China explained by rice versus wheat agriculture," *Sci-
    ence* 344, no. 6184 (2014): 603–608.
14. A. Alesina, P. Giuliano, and N. Nunn, "On the origins of gender roles: Women and the
    plough," *Quarterly Journal of Economics* 128, no. 2 (2013): 469–530.
15. A counterpoint to the narratives that environmental processes and factors from the dis-
    tant past are important for successful economic development in the modern era is the
    narrative of "reversals of fortune." According to it, not only are early fortunes not positive
    factors that predict modern development, if anything, they are *negative* factors. For ex-
    ample, Daron Acemoğlu, Simon Johnson, and James Robinson ("Reversal of fortune:
    Geography and institutions in the making of the modern world income distribution,"
    Quarterly Journal of Economics 117, no. 4 [2002]: 1231–1294) provide some evidence that
    geographic and environmental factors that made some societies rich in 1500 have actu-
    ally made these same areas poor in modern times. In their analysis, they focus on former

European colonies where some positive geographic factors (i.e., the presence of minerals) may have been related to European colonization, which, over time, served to make the native populations actually worse off because of the failure of institutions. More recently, even these reversal-of-fortune results have been complicated by new results from Areendam Chandra, C. Justin Cook, and Louis Putterman ("Persistence of fortune: Accounting for population movements, there was no post-Columbian reversal," *American Economic Journal: Macroeconomics* 6, no. 3 [2014]: 1–28), who show that migration patterns account for some of the "reversals." While the *places* that experienced positive environmental conditions in 1500 have had some reversals of fortune, the *peoples* who experienced these positive conditions in 1500 have instead had persistence of fortunes, because they (actually, their descendants) now live in other places than they did in 1500.

There is also a subfield emerging within the geohistorical viewpoint that adds further complexity to the snowballing effects of geographical advantages and disadvantages by examining whether the same geographic condition can have both positive and negative effects on later development—where the effects depend on still more factors. The "ruggedness" of local landscapes is a good example. Nathan Nunn and Diego Puga ("Ruggedness: The blessing of bad geography in Africa," *Review of Economics and Statistics* 94, no. 1 [2012]: 20–36) argue that ruggedness of the landscape can have opposite effects on country outcomes. It can have a negative cumulative direct effect on many aspects of modern-day country success by shaping factors such as agriculture, construction, and trade. Ruggedness makes it difficult to enclose farms, to travel to trade with neighbors, and to build sound homes and schools. Thus, mountainous regions within developed countries would lag behind flatter regions. However, there is a bright side to ruggedness. Consider Africa during the early days of the slave trade. Rugged terrain actually served as a barrier to slave raids and slave trading, allowing local populations to remain secure and build and expand villages and infrastructure while their neighbors in less rugged areas were brutally captured, transported to the New World, and enslaved. Indeed, research shows that the historical (indirect) positive effect of protecting populations may be twice as large as the contemporary (direct) negative effect.

16. This research is dredging up old controversies, many stemming from racist and xenophobic roots. As we have described in previous chapters, the integration of genetics and social science allows for both opportunities and pitfalls; combining population genetics, macroeconomics, and social science extends these issues. Indeed, new examinations of the intersection between genetics (and implicitly, race and ancestral origins) and national success and failure must recognize and reflect on (but not repeat) the historical use and misuse of science (and pseudoscience) in attributing the success of European countries and the lack of success of other (often African) countries to biological and genetic differences between "superior" and "inferior" "races" of peoples. As mentioned in chapter 5, many of these theories are being shown to be scientifically invalid (apart from their important ethical and moral problems) and, one hopes, will become relics of a less enlightened time. For example, in chapter 2 we discussed (and dismissed) *The Bell Curve*'s alarm about a coming genetic stratification (genotocracy). In chapter 5, we further debunked the "genetics of race." In addition, there exist the past misuses of previous "research" in genetics and biology to describe cross-national differences in outcomes, misuses that we want to acknowledge but move beyond. Such acknowledgement does not mean that the new integration of population genetics and social science will be uncontroversial or untainted by the past.

17. Q. Ashraf and O. Galor, "The 'Out of Africa' hypothesis, human genetic diversity, and comparative economic development," *American Economic Review* 103, no. 1 (2013): 1.

18. These calculations are actually more complicated than they sound, and there are several ways to create genetic diversity measures. In fact, we do not have sufficient genetic data from populations across the world that could be used to compare the genotypes of randomly drawn pairs of people in the same population space (e.g., country). That would require large samples of people from every country in the data set (109 countries are included in figure 6.1). Instead, data from a smaller set of 53 ethnic groups is used, sometimes with fewer than 100 people in a group. This information—the genetic diversity for, say, the Yakut ethnic group in Siberia, is calculated and then that number (the diversity score) is applied to any country in which Yakut people currently reside in proportion to their population—in this case, Russia. Another example is as follows. Researchers could estimate the genetic diversity of the French ethnic group and apply that number to the country of France (in line with the proportion of the population who is French, adding information for ethnic groups from North Africa and other places around the world). They would also apply the "French" level of diversity to the diversity score of the United States in proportion to the number of U.S. citizens who report having French ancestry, as well as to former French colonies that still have French populations. An alternative to using ethnic-group measures that are combined together to create country measures of diversity is to compare the relationship between the geographic location of each ethnic group to out-of-Africa migratory paths to construct predictions of genetic diversity for each country.

19. Ashraf and Galor, "The 'Out of Africa' hypothesis," 3.

20. A more complicated version of the theory is that genetic diversity allows greater task specialization. For example, there may be some goods that require both fine motor skills and gross motor skills to produce (i.e., the skills are complementary). Having both "types" of people could then allow goods to be produced that would otherwise not exist in places with only "fine motor" or only "gross motor" people.

21. E. Spolaore and R. Wacziarg, "Ancestry, language and culture," in *The Palgrave Handbook of Economics and Language*, ed. V. Ginsburgh and S. Weber (London: Palgrave Macmillan, 2015), 174–211.

    Yet another branch of macro-genoeconomics uses genetics to explain across-population interactions rather than patterns of growth. These studies take the logic of genetic diversity's pros and cons in explaining country development outcomes and ask how levels of genetic similarity (genetic distance) across countries determines patterns of relationships, like trade and violence. The macroeconomists Enrico Spolaore and Romain Wacziarg have been at the forefront of this research. In "War and relatedness" (working paper 15095, National Bureau of Economic Research, Cambridge, MA, June 2009, http://www.nber.org/papers/w15095), they showed that countries whose populations have higher genetic relatedness to one another engage in more interstate conflicts. This finding flies in the face of many theories of interstate and intergroup conflict. Dating back to the early 1900s, William Sumner formulated a hypothesis that ethnic dissimilarity would be associated with war and plunder whereas societies that are culturally similar would fight less; a second hypothesis in Spolaore and Wacziarg's article is that geographic—and not cultural—proximity is the key to understanding conflicts. It is difficult to unravel the effects of proximity from the effects of culture, since countries that share a border often share a history and culture to some degree. Indeed, the macroeconomists overturn these hypotheses by finding that even when controlling for geographic proximity, genetic similarity is tied to higher levels of conflict.

    Like the trade-offs for genetic diversity, Spolaore and Wacziarg outline potential trade-offs with genetic distance between countries. On the positive side, countries that are more genetically similar may share common cultural practices and ideals that could help them peacefully resolve conflicts. However, the downside of similarity and greater contact is the eventuality of disagreement and conflict, and some of the disagreements may be large enough to cause wars. Indeed, the authors' results suggest that the drawbacks outweigh the benefits with regard to wars. Because countries that are similar genetically tend to interact more, such patterns of interaction lead to more conflicts and (eventually) wars. It is a "contact" theory of conflict—countries get into conflicts with those with whom they interact and not those with whom they never have contact.

More recently, the same authors have proposed genetic distance between countries as a general measure (a summary statistic) of cultural similarity between nations. This is another important and controversial step in combining traditional social scientific measures of culture and norms with the explosion of genetic data from around the world. They show that measures of genetic distance across country populations are statistically related to other measures of "distance" between populations, such as linguistics, religion, and "values" taken from surveys about norms (i.e., having "traditional" family values or agreeing with notions of gender equality).

22. While not the most obvious choice of an outcome, population density is often used in country-level analysis during this period (pre-Columbian). In part, this represents our lack of data on gross domestic product in 1500 at the country-level around the world. Macro-economists claim that population density captures a broad set of conditions related to wealth and development, as only cities/countries with infrastructure and material wealth could have supported high densities over time.

23. Granting a cause-and-effect relationship is a big step in itself. How can Ashraf and Galor separate the potential role of other "lurking" processes that might be statistically associated with diversity and development and focus only on the cause-and-effect relationship between genetic diversity and country development? They propose an interesting approach to tease this out. They locate a variable—called an instrumental variable in social science research—that is linked with genetic diversity but not with the other processes that may affect development as well.

They pursue this approach by using a well-established finding from population genetics and historical analysis. The distance by foot from the cradle of human origins in East Africa's Rift Valley to a country predicts genetic diversity in that country. They claim that this measure of distance fills the role of an instrumental variable—that the variable only affects genetic diversity but not the other processes of concern. They also statistically control for many of the key alternative hypotheses in their analysis, such as colonization patterns, longitude and latitude of a country, dates of the Neolithic Revolution, and so on. Thus, they create this measure of geographic distance (by foot, out of Africa) for every country in their dataset and use the distance measures to form a statistical prediction for country-level genetic diversity. They then statistically link these levels of genetic diversity to country development. If their assumption is accurate—that migratory distances out of Africa are associated with country development only through its link with genetic diversity—then they can claim to have a cause-and-effect finding until someone else offers a plausible alternative explanation that correlates with both their genetic measure and development.

24. Alberto Alesina and Eliana La Ferrara review the literature that uses measures of ethnic diversity, fragmentation, and heterogeneity to examine aggregate economic outcomes ("Preferences for redistribution in the land of opportunities," *Journal of Public Economics* 89, no. 5 [2005]: 897–931). They also outline a broad array of evidence for the costs and benefits of this type of diversity, including country-level growth and the size and generosity of welfare states.

Combining the idea and empirical findings about the impacts of ethnic diversity on outcomes with the idea that ethnic diversity and genetic diversity are intertwined could then imply that measures of genetic diversity have an effect. However, stretching these findings from one area to another (even closely related) area requires additional support.

25. E. Spolaore and R. Wacziarg, "War and relatedness," *Review of Economics and Statistics*, 2016 (in press).
26. Lactase is the enzyme that breaks down lactose (a sugar that gives milk its sweet flavor). Without lactase, the consumption of dairy products can lead to lactose intolerance symptoms, such as abdominal cramping, diarrhea, etc. Mammals typically cease to produce lactase after weaning, but lactase persistence is unique to humans and evolved in the last 10,000 years. The *LCT* gene encodes the lactase enzyme, with variation coming from two SNP mutations. See J. T. Troelsen, "Adult-type hypolactasia and regulation of lactase expression," *Biochimica et Biophysica Acta* 1723, no. 1–3 (2005): 19–32.
27. C. J. Cook, "The role of lactase persistence in precolonial development," *Journal of Economic Growth* 19, no. 4 (2014): 369–406.
28. See also chapter 4, which discusses the advantages for mating partners of having different genetic variants related to immune system function.
29. Similarly, think about trying to defend your computer against viruses. Viruses are constantly finding holes in the standard computer virus protection software, and consumers are constantly upgrading their software. A community (i.e., an office) in which all have the same virus protection software may be more vulnerable to a given virus attack than a community that all has different software (and thus different vulnerabilities and strengths).
30. C. J. Cook, "The natural selection of infectious disease resistance and its effect on contemporary health," *Review of Economics and Statistics* 97, no. 4 (2015): 742–757.

CHAPTER 7: THE ENVIRONMENT STRIKES BACK: THE PROMISE AND PERILS OF PERSONALIZED POLICY

1. See the discussion in appendix 4 on epigenetics. Although epigenetic markers do not change the letters (e.g., CGC to CCC), they can turn off the gene's ability to form a protein. Since epigenetic markers are determined through environmental conditions, we must be cautious about treating the environment as an impediment and focusing only on the universal effects of genetic variants on outcomes.
2. While it is relatively uncontroversial nowadays to place nature and nurture on an equal footing in scientific inquiry, it has not always been this way, especially in the basic sciences. A history of discounting environmental factors strongly aligns with the history of genetic analysis more broadly. A century of heritability estimates that consistently pointed to genetics as a key source of variation in (nearly) all human behaviors and endeavors have, in some quarters, acted as blinders, obscuring consideration of any environmental factors (see chapter 2). The environment has largely been viewed as a nuisance in the task of understanding the genetic factors that lead to important outcomes (i.e., noise makes the genetic signal harder to detect).

   There are good reasons to blind oneself from the potential importance of the environment when attempting to uncover universal truths of biology and genetics. Blinders allow us to focus our attention on what we think is most important and prevent ourselves from being distracted by the complexity surrounding these truths. Clinical scientists and evolutionary scientists often believe (with good evidence) in the fundamental importance and invariance of genetics and its effects. Genes have specific biological functions— they code for proteins and contain the blueprint of human life and existence itself. It is necessary to understand how they work. To do this, scientists use methods that divorce gene function from environmental contexts and their potential effects. Scientists focus on the unchanging features of genes—functions that should always do the same thing.

More broadly, though, scientists often think of genes as unaffected by the environment, at least over short periods of time and in typical settings. This is a reasonable shortcut to take in modeling human behaviors because genes are stable over a human lifetime. Although the environment can shape human genetic profiles over generations by altering patterns of sexual reproduction and survivorship in a single generation, the human genetic code is too robust to be shaped meaningfully by the environment (save for exposure to massive amounts of ionizing radiation, as in the case of the Chernobyl nuclear disaster). That is, only about 30–100 new mutations appear in a typical person in each generation, often through errors during DNA replication. R. Acuna-Hidalgo, T. Bo, M. P. Kwint, M. van de Vorst, M. Pinelli, J. A. Veltman, A. Hoischen, L.E.L.M. Vissers, and C. Gilissen, "Post-zygotic point mutations are an underrecognized source of de novo genomic variation," *American Journal of Human Genetics* 97, no. 1 (2015): 67–74, http://www.cell.com/ajhg/fulltext/S0002-9297(15)00194-9.

However, new but preliminary evidence from the genetics research that used "deep sequencing" methods to uncover further variation in human DNA has begun to cast doubt on the sources of new mutations in the next generation. The classic location of new mutations is in the egg or sperm cell—so-called germline mutations—which are passed to the child. A recent discovery is a second source of mutation that appears postzygotically (after sperm and egg are combined): embryonic mosaicism. This fascinating discovery implies a state in which two or more genetically distinct cell populations in an individual develop from a single fertilized egg. Perhaps like a twin that "absorbs" its co-twin during the early phases of pregnancy, this mosaicism further disrupts our understanding of ourselves as humans with a single blueprint of DNA in every cell. Instead, it casts us as individuals with multiple blueprints—a veritable melting pot of DNA errors. The current evidence, however, is too new to have clear implications about the extent of the mosaicism in people. One study found that about 7 percent of DNA mutations are of the mosaic variety, although the study calculated this figure for a sample of only 50 people.

Thus, it is a reasonable approximation to treat a person's genetic structure as fixed and unaffected by the environment and focus instead on one direction of causation—from the genome to outcomes, rather than from outcomes (and environments and exposures) to the genome.

However, social scientists *do* care about how environments shape genes—not through Darwinian natural selection but through *interplay* and *interaction*. These types of interplay may allow the gene to produce societal advantages in some contexts and disadvantages in other contexts without actually affecting the functioning of the gene itself. It is how the environment and social context shape and transform biological functions into outcomes, from health and survival to fertility to economic success.

3. E. Turkheimer, A. Haley, M. Waldron, B. D'Onofrio, and I. I. Gottesman, "Socioeconomic status modifies heritability of IQ in young children," *Psychological Science* 14, no. 6 (2003): 623–628.

4. P. M. Blau and O. D. Duncan, *The American Occupational Structure* (New York: Free Press, 1978).

5. As does a famous early study of the effects of military service during the Vietnam War by the economist Joshua Angrist, which found that for whites in the 1980s there was a severe cost to being drafted that amounted to 15 percent of lifetime earnings whereas for blacks there was no wage penalty. The explanation was that if you faced a civilian labor market

with few opportunities for advancement, it did not make much of a difference if you lost a year or two of civilian experience due to the war. If you were confined to flipping burgers before the war, you were confined to flipping burgers after the war, and intervening military service would not make much of a difference. It should be noted, however, that later analysis by Angrist himself (with new coauthor Stacy Chen), as well as research by others, found that the apparent wage penalty for whites faded by the year 2000 (or was overestimated to begin with). J. D. Angrist, "Lifetime earnings and the Vietnam era draft lottery: Evidence from Social Security administrative records," *American Economic Review* (1990): 313–333; J. D. Angrist and S. H. Chen, "Schooling and the Vietnam-era GI Bill: Evidence from the draft lottery," *American Economic Journal: Applied Economics* (2011): 96–118.

6. We may also recall from chapter 3 that Guo and Stearns made a similar case about race. G. Guo and E. Stearns, "The social influences on the realization of genetic potential for intellectual development," *Social Forces* 80, no. 3 (2002): 881–910.

7. Recall that when there is positive genetic assortative mating, twin estimates underestimate heritability.

8. If we knew which part of the environment mattered, we could see if there was greater variation in that measure for lower-SES families than for higher-SES families, which would also yield this pattern of results. But we do not know which aspects of environment to measure.

9. Since these were white samples and the polygenic score is computed for whites only, we did not directly address the race question.

10. Of course, we are not capturing the lion's share of the genetic effect with this score. However, even if we had found an interaction effect with mom's education level (our measure of SES), we could not have ruled out that this was not, in fact, a gene-gene interaction. That is, it could have been that the factor raising or lowering the impact of the child's genotype was not mom's actual years of schooling, but her genotype that predicted her own years of schooling (e.g., her genetically based cognitive ability, for instance, that is proxied by her schooling).

11. In fact, the only variable that seemed to moderate the effect of the polygenic score (affect its effect) was the mother's own polygenic score. When a genotypically educationally advantaged child had a high-genotype mother, she was predicted to go farther in school than if she had been born to a genotypically average mother. In fact, this parental genetic measure was the only background variable that seemed to affect how much offspring genotype mattered. No putatively "social" variable had any influence on the genotype-phenotype relationship for offspring. The effect of the child's educational polygenic score did not change depending on the age of the parent, how educated the mom was, or the child's gender. The only factor that influenced the polygenic score effect was whether the mother herself had a high or a low polygenic score.

12. For a discussion of how the portion of variance explained by the polygenic score might relate to the overall genetic contribution to variation in phenotype, please see the appendix to D. Conley, B. W. Domingue, D. Cesarini, C. Dawes, C. A. Rietveld, and J. D. Boardman, "Is the effect of parental education on offspring biased or moderated by genotype?" *Sociological Science* 2 (2015): 82–105.

13. The issue is that in order for us to use a variant in the polygenic score, it must be linked to education more or less across all of the 54 data sets in our analysis.

14. J. Yang, R.J.F. Loos, J. E. Powell, S. E. Medland, E. K. Speliotes, D. I. Chasman, et al., "FTO genotype is associated with phenotypic variability of body mass index," *Nature* 490, no. 7419 (2012): 267–272.

15. See in particular appendix 4, which discusses findings from the field of epigenetics.
16. Leveraging these biological insights about the function of the MAO-A gene, some antidepressants and antianxiety medications—monoamine-oxidase inhibitors (MAOIs)—explicitly target this gene to reduce the activity of the gene. By doing so, the medications prevent the breakdown of monoamine neurotransmitters and thus increase their availability to apply the brakes to neurological stressors. These drugs have a long history. They were originally treatments for tuberculosis, but researchers noticed that depressed patients with tuberculosis seemed to grow less depressed (but still had other TB symptoms), leading to a monoamine theory of depression and wide adoption of MAOIs in the 1950s. Many of these drugs had negative side effects, which encouraged the shift to newer lines of antidepressant medications, though new MAOI drugs that have fewer side effects and interactions with food and other drugs have been approved in the last decade. Both MAOIs and the newer class—the SSRIs—demonstrate apparent paradoxes. Namely, in the candidate gene studies the reduced-activity versions of the genes were the ones that put individuals at risk for psychopathology, but the respective drugs to treat them seemed to work by suppressing the activity of the genes (i.e., mimicking the reduced-activity version of the gene). This paradox can be explained in several ways: (1) The candidate gene studies were wrong (we have already seen that they have not actually been replicated in any consistent manner). (2) The drugs act in a different way than we think they do, such as by stimulating brain-derived neurotrophic factor (BDNF) (to increase synaptic plasticity). (3) The brain has compensatory mechanisms that are the real effects.
17. In contrast, much research in this area uses retrospective accounts of childhood experiences gathered from adults, often coming years or decades after childhood. An issue with retrospective measures of childhood experience is that the reporting of the experiences may be shaped by events that have occurred since childhood. For example, individuals who as adults are not doing well may be more likely to report levels of childhood trauma as higher than they were "in fact." Adults who became particularly successful may "forget" about their past during the surveys. Thus adult success, or lack thereof, may *shape* reports of childhood trauma so that the statistical analysis linking childhood trauma with adult outcomes has the causal arrow going partially in the wrong direction.

    This type of reporting bias has a long history and affects many types of studies. Another example is provided by surveys that ask people about their current employment status and subsequently ask whether the person has any disabilities that limit their ability to hold a job. Some people who report that they are currently unemployed go on to "create" the presence of a disability in order to justify their employment status. These processes all fall under the umbrella of justification bias in surveys, that is, when people try to account for their current status by shaping responses to other questions. Thus unemployment status causes reports of disability to increase, and poor adult outcomes may cause some reports of childhood trauma that would not have been reported had the adult outcomes been more positive.
18. A. Caspi, K. Sugden, T. E. Moffitt, A. Taylor, I. W. Craig, H. Harrington, J. McClay, et al., "Influence of life stress on depression: Moderation by a polymorphism in the 5-HTT gene," *Science* 301, no. 5631 (2003): 386–389.
19. N. Risch, R. Herrell, T. Lehner, K.-Y. Liang, L. Eaves, J. Hoh, A. Griem, M. Kovacs, J. Ott, and K. R. Merikangas, "Interaction between the serotonin transporter gene (5-HTTLPR), stressful life events, and risk of depression: A meta-analysis," *JAMA* 301, no. 23 (2009): 2462–2471; K. Karg, M. Burmeister, K. Shedden, and S. Sen, "The serotonin transporter promoter variant (5-HTTLPR), stress, and depression meta-analysis revisited: Evidence of genetic moderation," *Archives of General Psychiatry* 68, no. 5 (2011): 444–454.

20. E. Walker, G. Downey, and A. Bergman, "The effects of parental psychopathology and maltreatment on child behavior: A test of the diathesis-stress model," *Child Development* 60, no. 1 (1989): 15–24, http://www.jstor.org/stable/1131067.

21. A novel direction in attempting to understand the reasons that "risky" variants may survive in the population through their role in being "beneficial" variants for other phenotypes is through a reversal of the data-mining of GWAS. Called PheWAS (Phenome-Wide Association Studies), researchers are sweeping through thousands of phenotypes against a single genotype to find novel associations for these putatively "risky" variants. See for example, M. Rastegar-Mojarad, Z. Ye, J. M. Kolesar, S. J. Hebbring, and S. M. Lin, "Opportunities for drug repositioning from phenome-wide association studies," *Nature Biotechnology* 33 (2015): 342–345. doi:10.1038/nbt.3183.

22. K. Donohue, L. Dorn, C. Griffith, E. Kim, A. Aguilera, C. R. Polisetty, and J. Schmitt, "The evolutionary ecology of seed germination of *Arabidopsis thaliana*: Variable natural selection on germination timing," *Evolution* 59, no. 4 (2005): 758–770, http://www.ncbi.nlm.nih.gov/pubmed/15926687.

23. S. F. Levy, N. Ziv, and M. L. Siegal, "Bet hedging in yeast by heterogeneous, age-correlated expression of a stress protectant," *PLoS Biology* 10, no. 5 (2012): e1001325, http://journals.plos.org/plosbiology/article?id=10.1371/journal.pbio.1001325.

24. For additional background discussion, see S. R. Jaffee and T. S. Price, "Gene–environment correlations: A review of the evidence and implications for prevention of mental illness," *Molecular Psychiatry* 12, no. 5 (2007): 432–442.

25. M. Rutter, "Gene–environment interdependence," *Developmental Science* 10, no. 1 (2007): 12–18, http://onlinelibrary.wiley.com/doi/10.1111/j.1467-7687.2007.00557.x/full.

26. A. Caspi, J. McClay, T. E. Moffitt, J. Mill, J. Martin, I. W. Craig, A. Taylor, and R. Poulton, "Role of genotype in the cycle of violence in maltreated children," *Science* 297, no. 5582 (2002): 851–854; A. Caspi et al., "Influence of life stress on depression."

27. C. Jencks, "Heredity, environment, and public policy reconsidered," *American Sociological Review* (1980): 723–736.

28. Even for the Jencks example, you might wonder how we could evaluate the claim that a genetic variant interacts with the environment (diet) to produce mental retardation. What if we suspected that the environment we are focusing on—diet—is actually partially determined by (passive) gene-environment correlation, in which families with resources are able to choose the diet and other families are not, and that the ability to choose the diet is (in some part) associated with genetic differences across families?

29. Take the example of schooling. Because the process of school assignment is not random and is probably correlated with genetic factors, efforts to estimate gene-environment interactions between school quality (environment) and a polygenic risk score face important challenges. One way to overcome these challenges is to focus on schooling environments that are assigned based on lotteries, test-score cutoffs, or experiments like the Tennessee STAR small-class-size intervention. J. B. Cullen, B. A. Jacob, and S. Levitt, "The effect of school choice on participants: Evidence from randomized lotteries," *Econometrica* 74, no. 5 (2006): 1191–1230; A. Abdulkadiroğlu, J. Angrist, and P. Pathak, "The elite illusion: Achievement effects at Boston and New York exam schools," *Econometrica* 82, no. 1 (2014): 137–196; A. B. Krueger and D. M. Whitmore, "The effect of attending a small class in the early grades on college-test taking and middle school test results: Evidence from Project STAR," *Economic Journal* 111, no. 468 (2001): 1–28.

30. Implicitly, we assume that genotype does not vary systematically by birth day. While some researchers have shown that genotype does vary by birth day across the year, the Vietnam lottery is assumed to break this association. See C. A. Rietveld and D. Webbink, "On the

genetic bias of the quarter of birth instrument," *Economics and Human Biology* 21 (2016): 137–146.

31. L. Schmitz and D. Conley, "The long-term consequences of Vietnam-era conscription and genotype on smoking behavior and health," *Behavior Genetics* 46, no. 1 (2016): 43–58.

32. J. M. Fletcher, "Enhancing the gene-environment interaction framework through a quasi-experimental research design: Evidence from differential responses to September 11," *Biodemography and Social Biology* 60, no. 1 (2014): 1–20, http://www.tandfonline.com/doi /abs/10.1080/19485565.2014.899454#.U2ZF-Bbo09U.

33. In this study there are four "groups" of people. One group has the risky genetic variant and was interviewed before 9/11; a second group has the risky variant and was interviewed after 9/11; a third group has the protective variant and was interviewed before 9/11; and a fourth group has the protective variant and was interviewed after 9/11. We use a difference-in-difference research design, which examines whether differences in depressive symptoms between the risky genotype group and the protective group following 9/11 are larger than the differences before 9/11.

34. Caspi et al., "Influence of life stress on depression."

35. Further adding to the intuition that the Caspi effect may be mis-specified was our study that used the randomness of birth-weight differences between identical twins as an environmental shock with which to interact *5-HTT*. Birth weight—as a proxy for prenatal nutrition and conditions in the womb more generally—has been known to predict all sorts of outcomes, from cognitive development to height to heart disease to wages. Since identical twins share their uterine environment, their gestational age, and their genes, the one with the lower birth weight can be thought of has having been randomly assigned to a calorie restriction by virtue of getting the back-row seat vis-à-vis the placenta. This approach—among others—has been used to ascertain the causal effect of fetal growth as distinct from other factors that may go along with it.

When we conducted this study and separated the twin pairs by *5-HTT* long-promoter genotype, we found that indeed those twins with at least one short (sensitive) allele were the ones who demonstrated a response to birth weight. However, our result was the opposite of what Caspi et al.'s diathesis-stress theory would suggest: additional nutrition raised the risk of later depression instead of lowering it! Of course, while the environmental stressor in this case is "randomly" assigned, genotype is not. Because we are comparing the effects of genotype across twin pairs (since monozygotic twins are by definition identical and thus we cannot look within the pair to see the effects of genotype), how do we know that it really is a gene-by-environment interaction and not an environment-by-environment effect? That is, perhaps we got the environment part right, and the short alleles are acting as a proxy for some environmental difference due to population stratification.

To address this issue, we analyzed dizygotic twins. These fraternal siblings shared uterine environment but are discordant on genotype. Thus, we can compare one twin that doesn't have any short *5-HTT* alleles with the sibling who has a short *5-HTT* promoter to see what the genetic effect is, free of any confounding with environment, since those differences (as discussed in chapter 3) are randomly assigned at conception. But we are then back to the problem of birth weight not being purely environmental, because the measured birth-weight difference between the twins is a result of both the randomness of implantation and placental architecture and their (differing) genetic predispositions to growth. It could, therefore, be a gene-gene interaction that we would be detecting: between the measured difference on the *5-HTT* promoter allele and some other unmeasured genotypic difference between the womb mates. The compromise in this dilemma was to run the analysis both ways: separately for the monozygotic and dizygotic twins, with the different assumptions embedded in each approach. If both analyses turned up the same answer, then we could be more sure—though still not 100 percent certain—that the approaches were converging on the "true" effect. The two approaches did indeed result in a strange, counter-Caspi finding. But mostly we mention this study to demonstrate the incredible challenges in getting at a truly causal gene-environment interaction. See D. Conley and E. Rauscher, "Genetic interactions with prenatal social environment effects on academic and behavioral outcomes," *Journal of Health and Social Behavior* 54, no. 1 (2013): 109–127.

36. J. D. Angrist and A. B. Krueger, "Does compulsory school attendance affect schooling and earnings?" *Quarterly Journal of Economics* 106, no. 4 (1991): 979–1014; J. M. Fletcher, "New evidence of the effects of education on health in the US: Compulsory schooling laws revisited," *Social Science & Medicine* 127 (2015): 101–107.

37. There are deep disciplinary differences in the importance of replication as a key ingredient for the believability of a study's findings. In genetic analysis, it is viewed as essential for a finding to be replicated in a second data set. In many social sciences, there is much less value placed on replication. These disciplinary differences may reflect divides about the importance of the environment and context in understanding causal processes. Geneticists may believe in processes that are invariant to context—gene A codes for protein B, regardless of the environment, whereas social scientists may have an overarching belief in context-dependency.

38. Our paper on the Vietnam lottery (Schmitz and Conley, "Long-term consequences") is an example of new research in gene-environment interaction work that uses polygenic scores rather than candidate genes. We think that the use of risk scores will become much more common in the near future because many data sets have shifted from having only candidate genes to being genome-wide genetic assessments.

39. In addition to using genome-wide data to form polygenic risk score measures, we can also use the data to statistically control for gene-environment correlation by deploying PC analyses (see chapter 2). Factoring out the effects of PCs to rid the data of population stratification gives us some clarity that our measured genetic effects are actual effects (rather than "chopsticks"), but such statistical legerdemain does not solve the problem that people choose their environments—perhaps based on the very genetic effects we seek to show their interaction with.

40. J. M. Donohue, E. R. Berndt, M. Rosenthal, A. M. Epstein, and R. G. Frank, "Effects of pharmaceutical promotion on adherence to the treatment guidelines for depression," *Medical Care* 42, no. 12 (2004): 1176–1185.

41. N. Tefft, "Mental health and employment: The SAD story." *Economics and Human Biology* 10, no. 3 (2012): 242–255.

42. E. A. Muth, J. T. Haskins, J. A. Moyer, G. E. Husbands, S. T. Nielsen, and E. B. Sigg, "Antidepressant biochemical profile of the novel bicyclic compound Wy-45,030, an ethyl cyclohexanol derivative," *Biochemical Pharmacology* 35, no. 24 (1986): 4493–4497.

43. A. Brayfield, ed., "Bupropion," in *Martindale: The Complete Drug Reference* (London, UK: Pharmaceutical Press, 2013), 107–111; L. P. Dwoskin, *Emerging Targets and Therapeutics in the Treatment of Psychostimulant Abuse* (Amsterdam: Elsevier Science, 2014), 177–216.

44. F. Chen, M. B. Larsen, C. Sánchez, and O. Wiborg, "The (S)-enantiomer of (R,S)-citalopram, increases inhibitor binding to the human serotonin transporter by an allosteric mechanism. Comparison with other serotonin transporter inhibitors," *European Neuropsychopharmacology* 15, no. 2 (2005): 193–198.

45. S. P. Hamilton, "The promise of psychiatric pharmacogenomics," *Biological Psychiatry* 77, no. 1 (2015): 29–35.

46. W. E. Evans and M. V. Relling, "Moving towards individualized medicine with pharmacogenomics," *Nature* 429, no. 6990 (2004): 464–468.

47. And there is more to come in treating tobacco use. One popular option is varenicline (Chantix), which is a nicotinic receptor partial agonist, meaning it stimulates nicotine receptors (nicotinic acetylcholine receptors) but not as strongly as nicotine itself stimulates them. By stimulating this receptor, it short-circuits the pleasurable effects of nicotine (similar to buprenorphine in heroin addicts) and attempts to reduce cravings. The targeting of specific genes opens the possibility that this treatment may be more effective in people who have specific variants of these genes. Another option for smoking cessation are the so-called nicotine replacement therapies (NRTs), for example, a nicotine patch or gum. Instead of short-circuiting the brain's response to nicotine, NRT feeds the brain nicotine in controlled doses as people try to quit by reducing withdrawal symptoms and cravings.

48. World Health Organization, *WHO Report on the Global Tobacco Epidemic 2008: The MPOWER Package* (Geneva, Switzerland: WHO, 2008).

49. We also briefly note the new and related field of nutrigenetics, which channels treatments toward individuals having genetic codes that can best take advantage of the treatments. Individuals are directed to the best "diet environments," which match their genetic predispositions for metabolizing fats, carbohydrates, etc.

50. J. M. Fletcher, "Why have tobacco control policies stalled? Using genetic moderation to examine policy impacts," *PloS One* 7, no. 12 (2012): e50576.

51. S. E. Black, P. J. Devereux, and K. Salvanes, "From the cradle to the labor market? The effect of birth weight on adult outcomes" (working paper w11796, National Bureau of Economic Research, Cambridge, MA, 2005).

52. See D. Conley and N. G. Bennett, "Is biology destiny? Birth weight and life chances," *American Sociological Review* 65, no. 3 (2000): 458–467.

53. A. Iliadou, S. Cnattingius, and P. Lichtenstein, "Low birthweight and Type 2 diabetes: A study on 11 162 Swedish twins," *International Journal of Epidemiology* 33, no. 5 (2004): 948–953, http://ije.oxfordjournals.org/content/33/5/948.short; J. Strohmaier, J. van Dongen, G. Willemsen, D. R. Nyholt, G. Zhu, V. Codd, B. Novakovic, et al., "Low birth weight in MZ twins discordant for birth weight is associated with shorter telomere length and lower IQ, but not anxiety/depression in later life," *Twin Research and Human Genetics* 18, no. 02 (2015): 198–209, http://journals.cambridge.org/action/displayAbstract?fromPage=online&aid=9657965&fileId=S1832427415000031.

54. C. J. Cook and J. M. Fletcher, "Understanding heterogeneity in the effects of birth weight on adult cognition and wages," *Journal of Health Economics* 41 (2015): 107–116.

55. Although it is difficult to evaluate the set of interventions because they are not randomly assigned to babies, economists have used a technique called a regression discontinuity design to attempt an evaluation. They leverage the common hospital practice of labeling babies weighing 5.4 pounds at birth as low birth weight, and babies weighing 5.6 pounds as normal birth weight. Thus, with a small difference in birth weight, a baby may receive a large difference in interventions, and we can ask whether these interventions are advantageous. Using the arbitrariness of the 5.5 cutoff for low birth weight, several economists have shown evidence that the interventions are relatively cost-ineffective and result in outcomes that are quite similar to babies who are born slightly above the 5.5 cutoff and therefore much less likely to receive the extra services and interventions. D. Almond, J. J. Doyle Jr., A. E. Kowalski, and H. Williams, "Estimating marginal returns to medical care: Evidence from at-risk newborns" (working paper w14522, National Bureau of Economic Research, Cambridge, MA, 2008). For follow-up discussion, see A. I. Barreca, M. Guldi, J. M. Lindo, and G. R. Waddell, "Saving babies? Revisiting the effect of very low birth weight classification," *Quarterly Journal of Economics* 126, no. 4 (2011): 2117–2123.

56. O. Thompson, "Economic background and educational attainment: The role of gene-environment interactions," *Journal of Human Resources* 49, no. 2 (2014): 263–294.

57. R. Haveman and B. Wolfe, "The determinants of children's attainments: A review of methods and findings," *Journal of Economic Literature* 33, no. 4 (1995): 1829–1878.

58. D. Lee, J. Brooks-Gunn, S. S. McLanahan, D. Notterman, and I. Garfinkel, "The Great Recession, genetic sensitivity, and maternal harsh parenting," *Proceedings of the National Academy of Sciences* 110, no. 34 (2013): 13780–13784, http://www.pnas.org/content/110/34/13780.short.

59. The Affordable Care Act (ACA; "Obamacare") explicitly forbids this type of cost-effectiveness analysis to drive decisions about coverage of procedures and treatments. But many economists describe this decision to ban cost-benefit analysis of specific medical interventions as equivalent to having written a blank check to pharmaceutical and medical device companies (among others), who will continue to produce very expensive new treatments even when the benefits are low. Future revisions of the ACA may shift to a NICE style model in order to control costs.

CONCLUSION: WHITHER GENOTOCRACY?

1. Or, "you probably will not enjoy eating asparagus."

2. In addition, they commit a common statistics sin by not also reporting that your probability (rather than odds) of, say, having a stroke is 1.2 percent (out of 100 percent) rather than 1 percent—a 20 percent increase in the odds but one displayed in a less hysterical way. N. Eriksson, J. M. Macpherson, J. Y. Tung, L. S. Hon, B. Naughton, S. Saxonov, L. Avey, et al., "Web-based, participant-driven studies yield novel genetic associations for common traits," *PLoS Genetics* 6, no. 6 (2010): e1000993.

3. On November 22, 2013, the FDA issued this ruling against 23andme's Personal Genome Service: "23andme, Inc.," Inspections, Compliance, Enforcement, and Criminal Investigations, U.S. Food and Drug Administration, Silver Spring, MD, http://www.fda.gov/iceci/enforcementactions/warningletters/2013/ucm376296.htm.

4. "FDA permits marketing of first direct-to-consumer genetic carrier test for Bloom syndrome," FDA News Release, U.S. Food and Drug Administration, Silver Spring, MD, February 23, 2015, http://www.fda.gov/NewsEvents/Newsroom/PressAnnouncements/ucm435003.htm.

5. Candidate gene assessments include *APOE* or *BRCA1/2* status, which can be useful in part because of the understanding that geneticists have about why these gene variants dramatically increase the risk of dementia or Alzheimer's and breast cancer, respectively.

6. For example, in the Add Health data, respondents born in June complete about two months less schooling by age 30 than those born in January.

7. http://www.babycenter.com.au/a1487/screening-for-down-syndrome.

8. *Gattaca* is a 1997 science fiction film starring Ethan Hawke and Uma Thurman that tells the tale of a dystopian society where genetic science has led to social engineering and "genoism."

9. Perhaps we will learn over time that this type of fetal selection will not matter very much, as society integrates the information, and parents and institutions adjust to compensate. If that is the case, an interesting implication of this potential waste of time and money is that it may lead to lower levels of inequality. This is because the rich will be wasting all the time and money and the poor will not (but may feel stressed-out from a sense of being left behind).

10. The theory is that, at least for male homosexuality, genotypes survive in the population because the sisters of gay men evince higher than average fertility, suggesting sexually antagonistic pleiotropy. R. C. Pillard and J. Michael Bailey, "Human sexual orientation has a heritable component," *Human Biology* (1998): 347–365.

11. See, for example, E. Telles, *Race in Another America: The Significance of Skin Color in Brazil* (Princeton, NJ: Princeton University Press, 2006). And on whites, see A. R. Branigan J. Freese, A. Patir, T. W. McDade, K. Liu, and C. I. Kiefe, "Skin color, sex, and educational attainment in the post-civil rights era," *Social Science Research* 42, no. 6 (2013): 1659–1674.

12. E. Oster, I. Shoulson, and E. Dorsey, "Limited life expectancy, human capital and health investments: Evidence from Huntington disease" (working paper w17931, National Bureau of Economic Research, Cambridge, MA, 2012), http://www.nber.org/papers/w17931.

13. "Scientists to sequence genomes of hundreds of newborns," *Nature* newsblog, November 23, 2015, http://blogs.nature.com/news/2013/09/scientists-to-sequence-hundreds-of-new borns-genomes.html.

14. C. Humphries, "Dating sites try adaptive matchmaking," *Technology Review* (2010), http://www.technologyreview.com/news/422216/dating-sites-try-adaptive-matchmaking/.

15. J. Streib, "Explanations of how love crosses class lines: Cultural complements and the case of cross-class marriages," *Sociological Forum* vol. 30, no. 1 (2015): 18–39.

16. When all the donors need to pass stringent requirements, we can be left with only five in an entire nation: http://www.telegraph.co.uk/news/health/news/11706863/UKs-national -sperm-bank-recruits-just-five-donors.html.

   For a full discussion of the reproductive cell market please see R. Almeling, *Sex Cells: The Medical Market for Eggs and Sperm* (Berkeley: University of California Press, 2011).

17. Yet another level of sorting is across platforms, where longer-term matches happen more on eHarmony and shorter-term matches happen more on Tinder.

18. B. W. Domingue, J. Fletcher, D. Conley, and J. D. Boardman, "Genetic and educational assortative mating among US adults," *Proceedings of the National Academy of Sciences* 111, no. 22 (2014): 7996–8000.

19. J. Price and K. Simon, "Patient education and the impact of new medical research," *Journal of Health Economics* 28, no. 6 (2009): 1166–1174.

20. This is called the fundamental cause hypothesis of health disparities. See, for example, B. B. Link and J. Phelan, "Social conditions as fundamental causes of disease," *Journal of Health and Social Behavior* (1995): 80–94.

21. See L. Schmitz and D. Conley, "The long-term consequences of Vietnam-era conscription and genotype on smoking behavior and health," *Behavior Genetics* 46, no. 1 (2016): 43–58.

22. J. R. Behrman, R. A. Pollak, and P. Taubman, *From Parent to Child: Intrahousehold Allocations and Intergenerational Relations in the United States* (Chicago: University of Chicago Press, 1995). Also see S. Marcus, "College education and the midcentury GI Bills," *Quarterly Journal of Economics* 118, no. 2 (2003): 671–708; and M. Page, "Father's education and children's human capital: Evidence from the World War II GI Bill" (working paper 06,33, Department of Economics, University of California, Davis, 2006).

23. L. L. Schmitz and D. Conley, "The impact of late-career job loss and genotype on body mass index" (working paper w22348, National Bureau of Economic Research, Cambridge, MA, 2016).

24. Though the populations are nearly always of European descent, which introduces another set of issues. How should we use polygenetic scores created from only white people on nonwhite populations? The predictions made from these scores on nonwhite populations are worse. The scores are not created using populations from nonwhite environments (e.g., Africa), so it is unclear how the score may interact with these environments. A similar issue has arisen in the drug discovery literature, where for decades only men were

used in clinical trials. Whether these drugs work the same (in dosage or even in the main effects) for women is unknown. More recently, the FDA, NIH, and other governing bodies have required that new medical technologies be tested on both men and women. No such requirement exists for genetic analysis across racial and ethnic groups.

25. J. Yang, R.J.F. Loos, J. E. Powell, S. E. Medland, E. K. Speliotes, D. I. Chasman, L. M. Rose, et al., "FTO genotype is associated with phenotypic variability of body mass index," *Nature* 490, no. 7419 (2012): 267–272.

   The results for each polymorphic locus are freely available online, so a plasticity score could be calculated by anyone who has their DNA code sequenced (through 23andme or otherwise). We, ourselves, have done the same using pedigree data that allows for a prediction of plasticity within families. D. Conley, B. Domingue, and M. Siegal, "Modeling the genetic architecture of phenotypic plasticity using sibling data" (working paper, Center for Genomics and Systems Biology, New York University, 2015).

26. L. Sweeney, A. Abu, and J. Winn, "Identifying participants in the Personal Genome Project by name" (white paper 1021–1, Harvard University, Data Privacy Lab, Cambridge, MA, April 24, 2013), http://www.forbes.com/sites/adamtanner/2013/04/25/harvard-professor-re -identifies-anonymous-volunteers-in-dna-study/#c39212d3e39f.

27. D. Lazer, *DNA and the Criminal Justice System: The Technology of Justice* (Cambridge, MA: MIT Press, 2004).

28. M. Jinek, K. Chylinski, I. Fonfara, M. Hauer, J. A. Doudna, and E. Charpentier, "A programmable dual-RNA-guided DNA endonuclease in adaptive bacterial immunity," *Science* 337, no. 6096 (2012): 816–821; D. Baltimore, P. Berg, M. Botchan, D. Carroll, R. A. Charo, G. Church, J. E. Corn, et al., "Biotechnology. A prudent path forward for genomic engineering and germline gene modification," *Science* 348, no. 6230 (2015): 36–38.

29. E. Lanphier, F. Urnov, S. E. Haecker, M. Werner, and J. Smolenski, "Don't edit the human germ line," *Nature* 519, no. 7544(2015): 410–411.

30. P. Liang, Y. Xu, X. Zhang, C. Ding, R. Huang, Z. Zhang, J. Lv, et al., "CRISPR/Cas9-mediated gene editing in human tripronuclear zygotes," *Protein Cell* 6, no. 5 (2015): 363–372.

31. Although the same issue exists for de novo mutations in each generation, mutations are rare and typically detrimental. Our discussion presumes an ability (in the future) for genome editing at a larger scale that is able to enhance fitness.

32. R. A. Sturm, D. L. Duffy, Z. Z. Zhao, F. P. N. Leite, M. S. Stark, N. K. Hayward, N. G. Martin, and G. W. Montgomery, "A single SNP in an evolutionary conserved region within intron 86 of the *HERC2* gene determines human blue-brown eye color," *American Journal of Human Genetics* 82, no. 2 (2008): 424–431.

33. A related issue is our lack of knowledge of all of the effects of each gene we might choose to edit. We often lack an answer to the question, why do humans have gene variant X if it is bad for us? One answer is that the genetic variant works in conjunction with other genes (a process called epistasis) to produce phenotypes, and if we edit the variant to reduce the likelihood a specific disease, the downstream effects on another phenotype may be devastating. If these downstream effects do not present themselves for many years, then the genetic edit could be truly disastrous and difficult (or impossible) to remedy.

34. J.-J. Rousseau, *On the Origin of Inequality*, reprint ed. (New York: Cosimo, 2005), 22.

35. Many other Enlightenment thinkers were more sympathetic to inequality than was Rousseau, who thought that the rise of private property was the primary culprit in all that is evil among men. The Scottish philosophers Andrew Ferguson and John Millar considered private property and inequality good for social progress, since they gave individuals something to strive for. One wonders, however, what they might think of inequality that has been (re-)written into the book of life. See, for example, A. Ferguson, *An Essay on the*

History of Civil Society, ed. F. Oz-Saltberger (Cambridge: Cambridge University Press, 1995); and J. Millar, *Observations Concerning the Distinction of Ranks in Society*, rev. 2nd ed. (London: 1773).

EPILOGUE: GENOTOCRACY RISING, 2117

1. Surrogacy had died out as a business because the epigenetic impacts of uterine environment (DNA methylation and histone acetylation) were seen as too important to be risked through outsourcing.
2. S. Cohn, C. Cohn, and A. Jensen, "Myopia and intelligence: A pleiotropic relationship?," *Human Genetics* 80, no. 1 (1988): 53–58, http://link.springer.com/article/10.1007/BF00451456.
3. E. M. Miller, "On the correlation of myopia and intelligence," *Genetic, Social, and General Psychology Monographs* 118, no. 4 (1992): 361–383, http://psycnet.apa.org/psycinfo/1993 –22443–001.
4. J. A. Driver, A. Beiser, R. Au, B. E. Kreger, G. L. Splansky, T. Kurth, D. P. Kiel, et al., "Inverse association between cancer and Alzheimer's disease: results from the Framingham Heart Study," *BMJ* 344 (2012): e1442, http://www.bmj.com/content/bmj/344/bmj.e1442.full.pdf.
5. C.M.A. Haworth, M. J. Wright, M. Luciano, N. G. Martin, E. J. de Geus, C. E. van Beijsterveldt, M. Bartels, et al., "The heritability of general cognitive ability increases linearly from childhood to young adulthood," *Molecular Psychiatry* 15, no. 11 (2010): 1112–1120.
6. C. A. Rietveld, S. E. Medland, J. Derringer, J. Yang, T. Esko, N. W. Martin, H.-J. Westra, et al., "GWAS of 126,559 individuals identifies genetic variants associated with educational attainment," *Science* 340, no. 6139 (2013): 1467–1471.

APPENDIX 2: A SECOND TRY AT REDUCING HERITABILITY ESTIMATES: USING GCTA AND PC METHODS

1. GCTA is the name of a common software package used to perform the analysis: J. Yang, S. H. Lee, M. E. Goddard, and P. M. Visscher, "GCTA: A tool for genome-wide complex trait analysis," *American Journal of Human Genetics* 88, no. 1 (2011): 76–82. GREML, meanwhile, stands for genomic-relatedness-matrix restricted maximum likelihood estimation.
2. We say approximately one-fourth because this figure can move up or down depending on the degree of assortative mating and the process by which parental education and residence is negotiated between the two parents.
3. S. M. Purcell, N. R. Wray, J. L. Stone, P. M. Visscher, M. C. O'Donovan, P. F. Sullivan, P. Sklar, et al., "Common polygenic variation contributes to risk of schizophrenia and bipolar disorder," *Nature* 460, no. 7256 (2009): 748–752.
4. A more recent and potentially more damning issue with GCTA has been presented by Kumar et al. (2016). The authors show evidence that the current use of only a subset of genetic variation in creating the genetic relatedness measures needed for GCTA analysis wreaks havoc on the whole approach. In much of social and biological science, incomplete measurement of an outcome or a predictor variable is a problem for analysis but is not catastrophic. For example, we regularly use SAT scores in the place of overall intelligence

outcomes in analyzing the impacts of schooling programs. Or we use a five-year average of annual family income reports to create measures of socioeconomic status to describe differences in the lives of children based on family resources. In each case, the measures are imperfect but are meant to reflect more general constructs of interest. But the particular methods used in GCTA require complete measurement of genetic relatedness. In cases where the true causal number of variants for a phenotype, say height, is in the thousands (or more), our current methods of measuring these variants may be too limited because most single-polynucleotide (SNP) chips cover only a small fraction of the genome. The authors show that this incomplete coverage of the causal variants can create extreme mismeasurement of genetically related values in the datasets typically used for GCTA analysis and that the extreme mismeasurement leads to heritability estimates that are often too big. Perhaps as whole-genome sequencing becomes the standard in the literature, these issues with GCTA can be reduced. S. K. Kumar, M. W. Feldman, D. H. Rehkopf, and S. Tuljapurkar, "Limitations of GCTA as a solution to the missing heritability problem," *Proceedings of the National Academy of Sciences* 113, no. 1 (2016): E61–E70.

APPENDIX 3: ANOTHER ATTEMPT—COMBINING PRINCIPAL COMPONENT ANALYSIS WITH FAMILY-BASED SAMPLES: A POTENTIAL WAY FORWARD (NOT YET ATTAINED)

1. Imagine that your Mom's genotype in a particular region is AG-CT; that is, she is heterozygous at two points on a given autosomal chromosome. You may get A and C from her while your brother—by chance—gets G and T at the two loci and your sister gets A and T. So, holding the contribution from your father constant for the moment, you would be 0 percent genetically similar (at these loci) to your brother and 50 percent similar to your sister. Do this for both chromosomes inherited from each parent and across the entire genome, and we get a measure of relatedness called identity by state for each pair of siblings.

APPENDIX 4: A TURN TO EPIGENETICS AND ITS POTENTIAL ROLE FOR MISSING HERITABILITY

1. These are but two regulatory regions. Others include enhancers, which can be thousands of base pairs away, but given the loops and bends in DNA when it is uncoiled into euchromatin (the active, transcribing state), they can touch the transcription machinery. The three-prime untranslated region (3′ UTR), that comes right after the end of the coding area also plays an important role in gene regulation through its interaction with micro-RNAs.
2. See figure "Chromosome 3 pairs" on webpage "Insights from identical twins," Genetic Science Learning Center, Univ. of Utah, http://learn.genetics.utah.edu/content/epigenetics/twins/.

3. X. Wang, D. C. Miller, R. Harman, D. F. Antczak, and A. G. Clark., "Paternally expressed genes predominate in the placenta," *Proceedings of the National Academy of Sciences* 110, no. 26 (2013): 10705–10710.

4. For a review see: H. A. Lawson, J. M. Cheverud, and J. B. Wolf, "Genomic imprinting and parent-of-origin effects on complex traits," *Nature Reviews Genetics* 14, no. 9 (2013): 609–617.

5. E. B. Keverne, "Genomic imprinting in the brain," *Current Opinion in Neurobiology* 7, no. 4 (1997): 463–468.

6. F. Ankel-Simons and J. M. Cummins, "Misconceptions about mitochondria and mammalian fertilization: Implications for theories on human evolution," *Proceedings of the National Academy of Sciences* 93, no. 24 (1996): 13859–13863.

7. See, e.g., J. Molinier, G. Ries, C. Zipfel, and B. Hohn, "Transgeneration memory of stress in plants," *Nature* 442, no. 7106 (2006): 1046–1049.

8. M. E. Pembrey, L. O. Bygren, G. Kaati, S. Edvinsson, K. Northstone, M. Sjöström, and J. Golding, "Sex-specific, male-line transgenerational responses in humans," *European Journal of Human Genetics* 14, no. 2 (2006): 159–166; G. Kaati, L. O. Bygren, M. Pembrey, and M. Sjöström, "Transgenerational response to nutrition, early life circumstances and longevity," *European Journal of Human Genetics* 15, no. 7 (2007): 784–790; M. E. Pembrey, "Male-line transgenerational responses in humans," *Human Fertility* 13, no. 4 (2010): 268–271.

9. See, for example, F. Torche, and K. Kleinhaus, "Prenatal stress, gestational age and secondary sex ratio: The sex-specific effects of exposure to a natural disaster in early pregnancy," *Human Reproduction* 27, no. 2 (2012): 558–567; R. Catalano, T. Bruckner, T. Hartig, and M. Ong., "Population stress and the Swedish sex ratio," *Paediatric and Perinatal Epidemiology* 19, no. 6 (2005): 413–420; R. Catalano, T. Bruckner, J. Gould, B. Eskenazi, and E. Anderson, "Sex ratios in California following the terrorist attacks of September 11, 2001," *Human Reproduction* 20, no. 5 (2005): 1221–1227; M. Fukuda, K. Fukuda, T. Shimizu, and H. Møller, "Decline in sex ratio at birth after Kobe earthquake," *Human Reproduction* 13, no. 8 (1998): 2321–2322.

APPENDIX 5: ENVIRONMENTAL INFLUENCES ON RACIAL INEQUALITY IN THE UNITED STATES

1. Of course, blacks are no longer the largest minority group in the United States, having been surpassed by Latinos, who as of 2014 make-up 17 percent of the population (https://www.census.gov/newsroom/facts-for-features/2015/cb15-ff18.html). But Latinos are a highly varied collection of ethnic groups who themselves display different admixed combinations and who also display a wide range of socioeconomic outcomes, even at the group level. So the black-white comparison is easier to make for illustrative purposes.

2. Blacks are about twice as likely to have a baby out of wedlock as whites. The nonmarital fertility rates among white and black women in 2010 was 32.9 and 65.3 per 1,000, respectively (Y. Kim and R. K. Raley, "Race-ethnic differences in the non-marital fertility rates in 2006–2010," *Population Research and Policy Review* 34, no. 1 [2015]: 141–159). Among Hispanics, the nonmarital fertility rate is about 80.6 per 1,000 (J. A. Martin, B. E. Hamilton, S. J. Ventura, M.J.K. Osterman, E. C. Wilson, and T. J. Matthews, "Births: Final data for 2010," in *National Vital Statistics Reports* 61, no. 1 [2012], http://www.cdc.gov/nchs/data

/nvsr/nvsr61/nvsr61_01.pdf). This difference is partly due to the huge differences in incarceration rates, which remove males from the marriage market. The per capita snapshot incarceration rate for whites is 446 per 100,000 (that is, the proportion of people under supervision by the criminal justice system at any given moment). For blacks it is a staggering 2,805 per 100,000. So at any given moment, six times as many blacks as whites are incarcerated at the state and federal level. This proportion does not include parolees; the Bureau of Justice Statistics compiles prison figures only; parole numbers are compiled from surveys and differ based on source (A. E. Carson, *Prisoners in 2013* [Washington, DC: Bureau of Justice Statistics, U.S. Department of Justice, 2014]). Whites also live longer than blacks do. Life expectancy at birth is 74.2 and 78.7 for non-Hispanic blacks and whites, respectively (E. Arias, "United States Life Tables, 2009," *National Vital Statistics Reports* 62, no. 7 [2014], http://www.cdc.gov/nchs/data/nvsr/nvsr62/nvsr62_07.pdf). The racial gap in life expectancy, however, has declined substantially in the last fifteen years. See G. Firebaugh, F. Acciai, A. J. Noah, C. Prather, and C. Nau, "Why the racial gap in life expectancy is declining in the United States," *Demographic Research* 31 (2014): 975–1006.

None of these statistics—and myriad others that we could cite—take into account gender differences. For example, black women earn 84 cents on the dollar of white women, while black men earn just 75 cents for each dollar that white men do. Median weekly earnings in 2013 for black and white women were $606.00 and $722.00, respectively. For black and white men, the figures are $664.00 and $884.00, respectively (Bureau of Labor Statistics, "Highlights of women's earnings in 2013," *BLS Reports*, no. 1051, December 2014, http://www.bls.gov/opub/reports/womens-earnings/archive/highlights-of-womens-earnings-in-2013.pdf). However, they give an overall flavor for the entrenched levels of inequality by race in U.S. society—even a half-century after the Civil Rights era that ended de jure segregation. It is that very persistence of racial disparities for generations that has led to fierce debate over their root causes. It was easy during the era of slavery, or Jim Crow that followed or even during the initial years after Civil Rights, to argue that the legacy of decades, centuries even, of ongoing, explicit racial oppression accounted for ongoing racial gaps, but why do they still persist today?

3. See D. Conley, *Being Black, Living in the Red: Race, Wealth and Social Policy in America* (Berkeley: University of California Press, 1999).

4. S. Fordham and John U. Ogbu, "Black students' school success: Coping with the "burden of 'acting white'," *Urban Review* 18, no. 3 (1986): 176–206.

5. M. S. Granovetter, "The strength of weak ties," *American Journal of Sociology* 78, no. 6 (1973): 1360–1380.

6. M. Bertrand and S. Mullainathan, "Are Emily and Greg more employable than Lakisha and Jamal? A field experiment on labor market discrimination," *American Economic Review* 94, no. 4 (2004): 991–1013.

7. Though often considered the gold standard, audit studies are not without their critics. For example, James Heckman has argued that audit studies may fail to eliminate all signals other than the intended ones through inadequate pairing of testers (J. Heckman, "Detecting discrimination," *Journal of Economic Perspectives* 12, no. 2 [1998]: 101–116). Second, respondents may be reacting to variance rather than mean differences even in correspondence studies (where face-to-face interactions are not used but rather mail or electronic communication). Recent work suggests that housing market discrimination is the most blatant and labor market bias is less robust to these concerns. See, for example, D. Neumark and J. Rich, "Do field experiments on labor and housing markets overstate discrimination? Re-examination of the evidence" (working paper w22278, National Bureau of Economic Research, Cambridge, MA, 2016).

8. D. N. Figlio, "Names, expectations and the black-white test score gap" (working paper 11195, National Bureau of Economic Research, Cambridge, MA, 2005).
9. A. R. McConnell and J. M. Leibold, "Relations among the Implicit Association Test, discriminatory behavior, and explicit measures of racial attitudes," *Journal of Experimental Social Psychology* 37, no. 5 (2001): 435–442.
10. R. Rosenthal and L. Jacobson, "Pygmalion in the classroom," *Urban Review* 3, no. 1 (1968): 16–20.
11. C. M. Steele and J. Aronson, "Stereotype threat and the intellectual test performance of African Americans," *Journal of Personality and Social Psychology* 69, no. 5 (1995): 797–811.
12. H. Schuman, *Racial Attitudes in America: Trends and Interpretations* (Cambridge, MA: Harvard University Press, 1997).
13. J. Citrin, D. P. Green, and D. O. Sears, "White reactions to black candidates: When does race matter?" *Public Opinion Quarterly* 54, no. 1 (1990): 74–96.
14. J. Hopkins, "No more wilder effect, never a Whitman effect: When and why polls mislead about black and female candidates," *Journal of Politics* 71, no. 3 (2009): 769–781.
15. J. G. Altonji and C. R. Pierret, "Employer learning and statistical discrimination" (working paper w6279, National Bureau of Economic Research, Cambridge, MA, 1997).
16. Sorting out cause and effect in situations with self-perpetuating dynamics such as these is quite difficult. Let us perform the heretical thought experiment of positing that Charles Murray and Richard Herrnstein were actually right when they wrote *The Bell Curve* over two decades ago. Certain "races"—namely, African Americans—lag on economic measures because they suffer from genetic "deficits" compared to their Euro-American counterparts with respect to cognitive ability. In this case, where the Herrnstein-Murray assumptions accurately reflect the real world, the racist attitudes and actions of whites (or blacks for that matter) might be well founded in economic efficiency. (Some diehard efficient-market worshippers might argue that if there really were equality in the productive abilities of underpaid, underemployed groups such as African Americans, savvy firms would rush in to profit from these "underpriced" workers and over time an equilibrium would be established in which black and white wages [and employment levels] should converge at least somewhat.) Disregard, for the moment, all the evidence that we discussed earlier that suggests environmental inequalities over the course of fetal, childhood, and adult development may give rise to real differences in ability that are wholly unrelated to genotypes. Herrnstein and Murray took IQ as a proxy for genetic ability. But, as we now know, those are hardly one and the same. IQ is an output and genotype is an input.

APPENDIX 6: IMPUTATION

1. "23andme Research Portal Platform," https://23andme.https.internapcdn.net/res/perma link/pdf/ashg/10292014_23andMeResearchPortal.pdf.
2. Granted, we randomly picked this SNP, and, as it turns out, if one uses the Japanese panel on 23andme, we find that it covers only one gene with a .3 r^2 threshold. But the larger point still holds.
3. B. W. Domingue, D. W. Belsky, D. Conley, K. M. Harris, and J. D. Boardman, "Polygenic influence on educational attainment," *AERA Open* 1, no. 3 (2015): 2332858415599972.
4. J. Daw (2014). Of Kin and Kidneys: Do Kinship Networks Contribute to Racial Disparities in Living Donor Kidney Transplantation? *Social Science & Medicine* (1982), 104, 42–47. http://doi.org/10.1016/j.socscimed.2013.11.043

INDEX

ACE model, 20–29; additive heritability in, 241n24; policy implications of, 29–34; variables in, 235n13. *See also* gene-environment interaction

Acemoğlu, Daron, 116–17, 127–28, 256n10, 256n15

Add Health. *See* National Longitudinal Survey of Adolescent to Adult Health

addiction, 39, 63, 161–62, 267n47

adoption studies, 14, 35, 72, 181, 218, 237n30

Affymetrix Corporation, 225, 230

African Americans, 14–15, 32–33, 62, 110–11, 219–24; civil rights movement and, 15, 274n2; ethnicity among, 89–90, 94; fertility rate of, 273n2; haplotypes among, 228; incarceration of, 182, 234n7, 274n2; Kwanzaa and, 251n9; principal component analysis of, 104–8; speech patterns of, 235n9. *See also* race

alcohol abuse, 12–13, 17, 160, 162

Alesina, Alberto, 259n24

Allen's Rule, 99

Alzheimer's disease, 42, 49, 173, 190

amphetamines, 39

Angrist, Joshua, 261n5

antidepressants, 39, 160–62, 239n6, 263n16

anxiety disorders, 35, 39, 40

apolipoprotein E (*APOE*) gene, 42, 49, 164; DNA screening for, 173, 175–76, 268n5

Arabidopsis thaliana (mustard weed), 149

Aronson, Joshua, 222

Ashraf, Quamrul, 8, 123–31, 134, 259n22

Asians, 7, 222; ancestry tree of, 87–88; chopstick problem and, 29; ethnicity among, 89; genetic diversity of, 8, 97, 105–6, 123–26; racial categorization of, 86, 185, 251n4

Asperger's syndrome, 186, 190, 193

assortative mating, 23, 204–5; cognitive genotypes and, 62; political beliefs and, 68–69, 247nn25–26; spouse selection and, 70–71

asthma, 50–51

attention deficit hyperactivity disorder (ADHD), 39, 41, 190, 193

audit studies, 274n7

Australopithecus, 90

autism, 186, 190, 193

back-breeding, 38, 217

Barker hypothesis, 238n31

BDNF gene, 165

Becker, Gary, 69

Bell Curve, The (Herrnstein and Murray), 2–5, 14–18, 34, 58–60, 109; economic measures and, 275n16; limitations of, 82–83, 225; propositions of, 60–66, 243n3; spouse selection and, 73–76, 249n38

Bendix, Reinhard, 242n1

BGI (Beijing Genomics Institute), 173

birth weight differences, 167, 267n55; IQ and, 163–66; in twins, 26, 265n35

Blackstone's Formulation, 53

Blau, Peter, 142

Blumenbach, Johann Friedrich, 88, 251n4, 251n6

Boardman, J. E., 243n8, 244n10, 245n12, 248n31

body mass index (BMI), 171; DNA screening for, 176; fertility and, 79, 80; GWAS on, 181; predicted standardized phenotype for, 64; spousal correlation with, 73–76